GARLING

# David Attenborough

## New Life Stories

# David Attenborough

## New Life Stories

Collins

BY THE SAME AUTHOR

*Zoo Quest to Guiana* (1956)
*Zoo Quest for a Dragon* (1957)
*Zoo Quest to Paraguay* (1959)
*Quest in Paradise* (1960)
*Zoo Quest to Madagascar* (1961)
*Quest Under Capricorn* (1963)

◇

*The Tribal Eye* (1976)
*The First Eden* (1987)

◇

*Life on Earth* (1979)
*The Living Planet* (1984)
*The Trials of Life* (1990)
*The Private Life of Plants* (1995)
*The Life of Birds* (1998)
*The Life of Mammals* (2002)
*Life in the Undergrowth* (2005)
*Life in Cold Blood* (2008)
*Life on Air* (new edition, 2009)
*Life Stories* (2009)

Published in 2011 by HarperCollins*Publishers* Ltd.
77-85 Fulham Palace Road
London W6 8JB

ISBN 978-0-00-742512-9

Frontispiece. *The wonder and variety of the natural world astounded the Victorians. John Gould, ornithologist and publisher, took particular delight in hummingbirds. He knew of – and in many cases, named – 418 species and published portraits of each of them in six sumptuous folio-sized volumes. This one, from the eastern Andes, he called the wire-crested thorntail.*



Printed in China by Toppan

# Contents


| | | |
|---|---|---|
| 1 | Canopy | 7 |
| 2 | Kiwi | 19 |
| 3 | Charnia | 29 |
| 4 | Foreign Fare | 41 |
| 5 | Cicada | 51 |
| 6 | Earthworms | 61 |
| 7 | Wallace | 69 |
| 8 | Hummers | 81 |
| 9 | Identities | 93 |
| 10 | Rats | 105 |
| 11 | Monsters | 115 |
| 12 | Butterflies | 125 |
| 13 | Chimps | 137 |
| 14 | Cuckoo | 147 |
| 15 | Quetzalcoatlus | 157 |
| 16 | Chameleon | 167 |
| 17 | Nectar | 179 |
| 18 | Waterton | 191 |
| 19 | Fireflies | 201 |
| 20 | Elsa | 211 |
| Acknowledgments | | 220 |
| Index | | 222 |

# 1

# Canopy

Fifty years ago – *only* fifty years ago – one part of the tropical rain-forest was virtually unexplored. Unknown species certainly lived there; yet few scientists had gone there to look for them.

◁ *A long jib-arm, swivelling on the top of a 120 foot tall tower in the Venezuelan rain-forest, enables botanists – and film-makers – to inspect every tree in the canopy below.*

Fifty years ago – *only* fifty years ago – one part of the tropical rain-forest was virtually unexplored. Unknown species certainly lived there; yet few scientists had gone there to look for them.

Botanists, in particular, had a special reason for getting there. The trunks of many tropical trees look virtually identical. Smooth and cylindrical like the columns of a Norman church, they rise for fifty to a hundred feet before they produce a single branch. Distinguishing one species from another can be extraordinarily difficult, sometimes virtually impossible.

A patch of brilliantly coloured blooms lying on the forest floor can give a clue. But which tree have they come from? It is difficult, sometimes impossible to be certain.

Back in the late 1940's, a botanist, later to become a Professor at Cambridge, found a solution to the problem. Monkeys – pig-tailed macaques, to be precise. Local people in Malaysia trained them to clamber up coconut palms, twist off the nuts and throw them down to the ground. Professor Corner acquired one and trained it to break off the fruit or the flowers of any tree he indicated. So a significant proportion of the Malaysian flora written immediately after the last war was compiled – initially at any rate – by a monkey.

Monkeys might have helped botanists, but they were of little use to zoologists who wanted to know about the animals that lived up there. So one or two of the wealthier scientific institutions started to build towers tall enough to project above the canopy, with a jib arm at the top like those on cranes that are used to build sky-scrapers.

Back in the 1990's I was making a series of programmes about birds and I wanted to film those that live in the canopy. We discovered that one of these giant towers had been erected in the middle of the Venezuelan rain-forest around the headwaters of the Orinoco by an Austrian scientific institution. It sounded ideal.

We got all the necessary permissions and off we went. It wasn't easy to get to the site. We had to charter a small plane to take us to a remote mission station. From there we had a two-hour trip in canoes down the Orinoco and then an hour or so of hiking through the forest. But eventually we reached the scientists' base – a little cluster of small huts with beds and one large building which contained the canteen and laboratories. A bearded scientist was sitting at the dining table as we walked in, reading a book and puffing a pipe. We introduced ourselves, explaining that we were making a series of films about birds.

'Birds? ' he said. He couldn't have sounded more astonished than if I had said we had come looking for fairies.

'Vye do you come here for *birds?*'

'Well, you do have some, don't you?' I said.

'I go up zee tower every day,' he said, ' And in zee last three weeks I have seen one. But not more.' And with that he shut his book and went out.

We were baffled. Over a thousand different species of birds have been recorded in Venezuela. But alas! our Viennese host – he proved to be a botanist – was only too correct. The tower was made of steel girders and stood a hundred and twenty feet tall. In order to get this gigantic structure into this remote place, the scientists recruited the aid of a gigantic Russian-made, twin-rotor helicopter, operated by the Venezuelan Air Force. It had lifted the sections of the tower from the nearest stretch of metalled road, carried them across many miles of the forest, and lowered them into place, one by one. But the force of the down-draught from the helicopter's two huge rotors had blown off every flower, every fruit and every epiphyte from the forest branches beneath. So now there was nothing to tempt a bird to live there. The scientist was absolutely right. We went up there before dawn, day after day – but not a single bird appeared. So when it comes to filming the canopy, towers are not necessarily as helpful as you might think.

As time went by, we tried less large-scale solutions. You can use ropes. First you shoot a weight of some kind, with a long piece of string attached to it, over a high branch of one of the trees. Once you have done that, you can use the string to haul up a climbing rope strong enough to support the weight of a human body.

From then on, you have a choice of methods, most of which have been developed by rock climbers. The simplest is to use a pair of jumars. These are handles that you clip one above the other on to the rope up which you are climbing. Each has a rope loop hanging from it into which you put your foot. Standing with your weight on the lower one and bending your other knee, you hitch up the upper handle eighteen inches or so. Then you straighten that knee and up you go. It sounds easy.

As far as I know, it is the most exhausting way of rising vertically known to mankind. The first time I did it, I laboured away for what seemed to be about ten minutes. Sweat cascaded from me but nonetheless feeling rather proud of my progress, I looked down and discovered that I was only about six feet above the ground.

If and when you manage to summon up enough strength to get to the top, you are then faced with the question of how to get down again. It is especially difficult if you don't in fact clamber on the branch from which you are suspended but have to make the necessary adjustments to the jumars while you are still clinging to the rope. To do that you have to use a special knot to secure yourself and then change the setting on the handles. When the time came to do so, I repeated my instructions out loud to myself, interlaced I dare say with a few expletives. However, I got it done eventually and by the time I had slipped gracefully down the rope and landed at the bottom I had recovered something of my composure.

Dickie, the recordist, asked me how I had got on up there.

'A piece of cake,' I said casually.

'Well, that's not what it sounded like to me'. He said. 'You had better listen to what you said.'

And I had fallen into the old trap of forgetting that I had been fitted with a remote radio-mike – and Dickie had recorded my entire panic-stricken soliloquy.

The technique served me quite well for several more trips into the canopy. But then, as it was getting more and more difficult, I came to the conclusion that my rope-climbing days were coming to an end. And at that point, I met Phil, a superb naturalist and bird enthusiast and not only an expert rope-climber but one of the strongest men I have ever met. He devised a splendid way of getting me up to the canopy that involved pulleys and a counterweight. He filled a kit-bag with rocks and earth so that it was the equivalent of my weight and then using the pulley, hauled it up so that it hung just above the branch on which I was to sit. I attached myself to the bottom of this rope and, on a given signal, Phil made certain adjustments, checked the safety rope and then released the kit bag so that as it descended, I rose sedately – and indeed effortlessly – upwards.

The first time I used it was in New Guinea. I wanted to watch a group of greater birds of paradise performing their dances. The tree in which they displayed was an emergent – a particularly tall giant tree that rose above the surface of the canopy and stood almost alone. I was in another emergent, about twenty yards away. As I hung beneath my branch, I had a clear unimpeded view of the birds. It was a sight I shall not forget. The crowns of the lower, more numerous trees beneath me were pressed closely to one another, so that the leafy surface was continuous, like the surface of a cauliflower. The birds were about twenty yards away displaying in the sunshine, oblivious of my presence. Each one danced by lowering his head and erecting his golden plumes into a glistening fountain, shrieking in ecstasy at the appearance of a female.

I stayed up there, oblivious of my height above the ground and the patience of Phil who was masterminding the rope controls on the ground beneath. But eventually the time came to return. I called down to Phil – and he slowly lowered me down. The only fly in the otherwise unforgettable ointment was when, exactly half way down, I passed my counterweight kit bag on its way up. It was grotesquely huge – and I realised, with regret, that this probably was my last visit to that wonderland, the canopy.

*Nineteenth-century naturalists seldom saw the inhabitants of the rain-forest canopy unless they came down to attack them, as happened to Henry Bates in Brazil.*

Right. *The least exhausting way to reach the forest canopy is to allow others to hang a rope from the branch of one of the tallest trees and then set up a system of counterweights to pull you up. The arrangement here enabled me to watch birds of paradise displaying in New Guinea.*

Below. *In Malaysia, people train pig-tailed macaque monkeys to twist off coconuts and let them fall to the ground, though only the most trusted monkeys are allowed off a leash to do the job.*

*Macaws are big birds, with wings about three feet across, so they prefer to make their journeys in the open sky either along rivers or above the forest canopy. Here a small flock is probably on its way to gather food from a tree that has just come in to fruit. At other times, mated pairs tend to travel together, flying so close to one another that their wings seem almost to touch.*

*Spider monkeys are the most accomplished climbers in the canopy of the South American rain-forest, for their long prehensile tail can grip a branch as securely as can either their feet or their hands.*

Right. *Toucans, in spite of their slightly comic appearance, can be quite ferocious, as Henry Bates discovered. Unlike many of the birds in the canopy, they are meat-eaters, and will take chicks from the nests of other birds.*

Below. *Kangaroos might seem to be the most unlikely of creatures to take to a life high in the trees and perhaps it was only the absence of large tree-living mammals such as monkeys in the forests of northern Australia and New Guinea that lured them up there. Even so, they must be the clumsiest of all tree-livers and have been known, when crashing away in fright, to lose their grip altogether and fall to the ground.*

# 2

# Kiwi

If you wanted to pick a bird that least resembles a bird, you could do no better than pick the kiwi. Its feathers are not broad with barbs that zip together on either side of a central quill, like normal feathers. They are long and so thin that they are more like stout hairs.

◁ *A kiwi may have several burrows in its territory in which it spends the day, only venturing out about an hour after sunset. A burrow may be as much as 20 metres long, with a chamber at the far end big enough to accommodate two birds. Most unusually for a bird, kiwis mark their ownership of a burrow with strongly smelling droppings.*

If you wanted to pick a bird that least resembles a bird, you could do no better than pick the kiwi. Its feathers are not broad with barbs that zip together on either side of a central quill, like normal feathers. They are long and so thin that they are more like stout hairs. And they grow so densely that their owner seems to have not plumage but a shaggy coat like a dog's. Its wings? You have to look hard to find them. They are mere stumps, only a couple of inches long and normally invisible, buried deep in its coat. Its tail? It doesn't have one. And as far as the general outline of its body is concerned you might well say that this is a bird that has gone pear-shaped.

Kiwis spend their days in burrows which they dig for themselves and only come out at night. It is true that a few other birds, like owls and shearwaters, also do the same, but whereas those birds have extremely acute night vision, the kiwi's eyesight is very poor and it makes its way through the darkness relying more on its sense of smell than its sight. And in a most un-bird-like way, it actually marks its territory with its smelly droppings.

There are three species of kiwi, all chicken-sized and very much the same, and they live only in New Zealand. I once went down to the South Island to try and film them on a remote and lonely beach. They lived in the thick forest at the head of the beach, but I was told that sometimes they came out on to the sand to look for sand hoppers and that then we would be able to get a clear view of them. So I lay down on a strand line and draped some rotting seaweed over my shoulders. I reckoned it smelt strong enough to conceal any odour that might come from me.

About an hour after sunset, a small hunchbacked shape appeared at the head of the beach. It stood there, with its long beak touching the ground as though it was a tripod. Then it strolled slowly down to the strand line and started turning over the weed with its long beak and probing the sand beneath for the hoppers. Uniquely among birds, a kiwi's nostrils are not at the base of its beak but at the tip. That must be very useful when it needs to sniff out small creatures in leaf litter. Now however, it had to blow down its nose every few seconds to clear its nostrils of sand.

Puffing and snorting, it waddled slowly towards me, along the strand line, probing as it came. When it was about eighteen inches away, it looked up and gazed at me with its small bead-like eyes in a baffled sort of way – and then slowly walked round me and carried on with its probings. If there were to be such a thing as a hobgoblin, my guess is that it would look something like that strange little being. It certainly didn't seem to be much like a bird.

Why should it be so odd? Well, about 80 million years ago, the islands of New Zealand were connected to mainland Australia. At that time, the dinosaurs were approaching the end of their dominance of the earth. There were a few shrew-like mammals around, but they had only evolved relatively recently, in geological terms, and were still tiny and insignificant, kept in their place, no doubt, by the dinosaurs. But the birds were flourishing. There were lots of different and greatly varied species but one group of them in particular had made a very bold move. They had abandoned flight and grown large and powerful as if they were making a bid to take over the land from the dinosaurs. And they did pretty well. In fact, they survived whatever catastrophe it was that exterminated the dinosaurs and their descendants are still around to this day.

They are the ostriches of Africa, the rheas of South America and the emus of Australia. The ostrich, the biggest of them, is taller than a man and has a kick that can rip open an animal's stomach. And there were some of these flightless birds living in New Zealand when it drifted apart from Australia. They were the moas. Some were about the size of turkeys, but one species was the tallest bird that has ever lived, standing around nine feet tall. It must have been very formidable indeed.

These huge creatures dominated the land of New Zealand for there were no mammals of any kind on the islands. What is more they continued to do so until human beings – the Polynesians – arrived on the islands about a thousand years ago. Within a few centuries the human settlers had exterminated them. Only their small secretive forest-loving cousins, the kiwis, remain.

There were, of course, plenty of other birds in the New Zealand forests that *could* fly. But flying is a very energy-demanding business. If you don't have to, why do so? There were no mammals here to prey on them. No rats or weasels, no dogs or cats. So some of New Zealand's flying birds took to spending most of their time on the ground and eventually gave up flying altogether.

One of the first to do so was a member of the parrot family, the kakapo. It was a leaf-eater and fed on grasses and in low bushes, plodding around from one to another, clambering about in the branches looking for new tender shoots. If you don't fly, there is little need to keep your weight down. And the kakapo didn't. In fact it became the largest and heaviest of all parrots, measuring a good two feet long from head to tail. It is now entirely ground-dwelling and ambles around in its territory along well-worn paths, nibbling leaves. One of the island's water birds, a kind of coot called the takahe, also took a similar line and abandoned flight. And now it is the giant of all coots, almost as big as a goose.

Birds, of course, aren't the only creatures that fly. Insects do so as well. And some of New Zealand's insects reacted to their isolation in exactly the same way as the birds did. The wetas belong to the grasshopper family. They too lost their wings and became giants. The biggest of them is now the size of a mouse. If you try to pick up one of them, you have to keep your nerve. First it brings forward its long back legs, armed with spines which it can thrash forward with great force. And then it opens its jaws and threatens you. A weta bite can easily draw a lot of blood.

Some mammals too, of course, evolved wings. They were the bats. They developed long after New Zealand had split away from Australia, but since they could fly, they were able to cross the arm of the sea that separated the two lands. And one of them, the short-tailed bat, is now also starting down the road to flightlessness. It seems to be a comparatively recent move for it is still very capable in the air and it still roosts in the holes in trees.

But it also digs holes in the ground. Indeed, it is the only bat in the world to do such a thing. Its back legs, which in most bat species are little more than suspension hooks, are unusually long and well-muscled. And its front legs – its arms, as it were – have a joint at the base of the elongated middle finger which supports the wing membrane. This allows that finger to be folded right back, so furling the wing which then slots into a groove on the lower arm. So the New Zealand short-tailed bat walks on its wrists and can and often does run around four-footedly, nuzzling through the leaf litter using its tiny snout to search for small worms and insects.

So one way and another, the restricted somewhat arbitrary selection of animals that happened by accident or intention to be living in New Zealand when it separated from Australia, have between them evolved to fill all the various niches that in the rest of the world are now occupied by mammals. The tallest of the moas browsed on the branches of trees, like giraffes. Smaller kinds cropped the bushes as goats do. A parrot, the kakapo, has turned, in effect, into a rabbit and the short-tailed bat is now taking up the way of life of a shrew.

And what about the kiwi? Well, it lives in holes in the ground which it digs for itself; it eats all kinds of small soil-living creatures but is particularly fond of earthworms; its feathers are like a coarse fur; it has poor eyesight but an acute sense of smell; and it stakes out its territory with piles of droppings. So it is not so much a bird – as a badger.

*The first printed illustration of a kiwi, published in London in 1813. The artist had never seen a living kiwi and had to base his illustration on a shrivelled skin that arrived in the British Museum in that year. He can hardly be blamed for consciously or unconsciously basing his reconstruction on the only other flightless and similar-sized birds with which he would have been familiar, the penguins. The kiwi was, in fact, regarded for some years as their close relation, though actually it belongs to the group that contains the ostrich and the extinct moas.*

Right. *Moas, now extinct, once dominated the forests and plains of New Zealand. There were a dozen or so species, ranging from the size of a turkey to the biggest, of which this is a model. It was the tallest bird ever to have existed and may have stood about nine feet high. The Maori had no knowledge of the origin of the huge bird bones that were found in their islands. Accordingly, Europeans named the birds moas, which is a commonly used term in the Polynesian Pacific for, rather unflatteringly, chickens.*

Below. *Kiwis are closely related to the extinct moas. Being shy, nocturnal and living in dense forest, they are seldom seen in the wild. But on one remote beach they venture into the open in order to probe into the sand in search of hoppers.*

Above. *The takahe, a giant flightless coot three times as heavy as its European relatives, had no defence against New Zealand's introduced predators and was considered extinct in the 1930s. In 1948, however, it was rediscovered grazing the high alpine slopes of South Island.*

Below. *The kakapo is a parrot. It too has abandoned flight and become a giant. By 1994 only eight proven fertile females survived but they were taken to predator-free offshore islands and in 1997 they started to breed once more.*

Right. *Flying could be dangerous for insects living on small islands. A sudden gale could sweep a whole population out to sea. New Zealand's wetas are grasshoppers that have given up flight. Since there is no pressure to remain small, wetas have become very big and powerful. They are also ferocious. This one has raised one of its heavily spiked hind legs in a threatening way and will thrash it and the other of its pair with such power that it can easily draw blood.*

Right. *The short-tailed bat is New Zealand's only native mammal. It now regularly hunts on the ground, running about in a most agile way on its hind feet and front elbows. Perhaps it has started down the road to flightlessness.*

# 3

# Charnia

Charles Darwin had one great problem when assembling his history of life on this planet. He knew nothing about its very early stages.

◁ Charnia masoni, *the first fossil to be found in rocks of undisputed Precambrian age. Seven inches long, it was discovered in the rocks of the Charnwood Forest.*

Charles Darwin had one great problem when assembling his history of life on this planet. He knew nothing about its very early stages. He certainly didn't know how one inorganic molecule became a living one – we are still wrestling with that problem – but he also didn't know what were the very first organisms that anyone might recognise as an animal or a plant.

The place to look for answers to that question, of course is in the rocks. In Darwin's time, the oldest fossils anyone had found – or at least recognised – came from rocks belonging to a period called the Cambrian, which we now know to have been deposited around 540 million years ago. But these Cambrian fossils are remarkably complex. Some are shells, like small clams. Others – trilobites – look like the woodlice that we find in our gardens – except that these were sometimes as big as rabbits. But who could believe that creatures with complex eyes, feelers and a dozen or so jointed legs beneath a coat of shelly armour like a trilobite were the very first animals that ever existed?

There were a number of possible explanations, of course, of this lack of simpler fossils. Maybe animals before then did not have any hard protection such as shells or armour but had naked jelly-like bodies so that when they died, they simply dissolved without leaving any trace. Or maybe the chemistry of the seas in those far-distant times was such that it was not possible for an animal to extract the chemical compounds needed to build a hard shell. Or maybe it was simply that all the rocks that date from that unimaginably ancient period have been so compressed, buckled, baked and mangled by the earth's geological processes that they could not possibly retain any trace of an animal's existence.

At any rate, one thing was clear to Darwin and other scientists of his time – the rocks that lay beneath the Cambrian rocks – the so-called Precambrian – were without fossils of any kind. So ideas about the very first chapter in the story of life could only be speculation. And that was still the situation when I was a boy, back in the 1930's. I lived in Leicester, in the middle of England, and I was an enthusiastic collector of fossils. To the east of the city, there were quarries from which came iron-rich Jurassic limestone. It was formed around 175 million years ago and it was full of spectacularly beautiful fossils – coiled shells like ammonites, and things like bullets – belemnites – that were once part of the internal structure of squid-like animals. There were even vertebrae of marine reptiles – ichthyosaurs – that swam around in those Jurassic seas, though I was never lucky enough to find one.

But the rocks to the north-west of the city were of no interest to me. They were, it is true, very ancient. Indeed, they were Precambrian. But that meant that – by definition – they didn't contain any fossils. So I didn't waste my time looking there.

How misguided I was. A mere eleven years after I had left my Leicester grammar school, a boy from the very same school, named Roger Mason, who was also interested in fossils, was playing with a couple of friends in a quarry in Charnwood Forest. The rocks there are composed largely of thick deposits of ash from ancient volcanoes that now have become compressed and very hard. Half-way up the face of the quarry, imprinted on the rock, he saw what looked like the impression of a frond of bracken. It was about seven inches long, with a series of branches sprouting on either side of a central rib. He told his father, a local minister who in turn persuaded a geologist on the staff of Leicester University to go and have a look at his son's find. There could be no doubt. It was a fossil – and it was in the Precambrian. It was named *Charnia masoni*; *Charnia* after where it had been found and *masoni* after the boy who had found it. And it caused a geological sensation.

And not least on the other side of the world in Australia. There, ten years earlier in the Ediacara hills north of Adelaide, geologists had found very similar fossils. No one at the time had been able to ascertain the age of the Ediacaran rocks. They were certainly extremely ancient, but people argued that the very fact that they contained fossils meant that they must be Cambrian. Now the Leicestershire find, in rocks that were undoubtedly Precambrian led the Australians to change their ideas. And fossil hunters world-wide began to look more carefully and intensively in rocks that were so old they had assumed hitherto that they were devoid of life.

In most areas that remained true. They *were* without fossils of any kind. But in just one or two places conditions on the sea-floor had been like those when the Charnwood rocks were laid down. The animals living there had been buried under layers of volcanic ash that had preserved an impression of their dead bodies – and the rocks that contained them, unlike so many of their great age, had remained largely unaffected by the titanic movements in the earth's crust that had occurred in the millennia that followed.

Last year, I went to one of those very special places – Mistaken Point that lies on the east coast of Newfoundland. It is called 'Mistaken' because in the old days, ships sailing through the fogs that so often shroud this part of the world, mistakenly thought that they had cleared land and were heading for the open sea – only to crash into the rocks of the Point.

On the day we were there, the weather was glorious. And the fossils were amazing. There are great flat areas of rock that though they are now tilted were once the horizontal floor of a sea that had been covered with a thick layer of volcanic ash. And in one place, on a flat surface about the size of a tennis court, there were, to my astonishment, Charnias. Not one. Hundreds of them, lying flat and roughly parallel to one another. It seemed that ash in the water had swept in from one particular direction so that they were all more or less aligned. It needed only a little imagination to conjure up a vision of that ancient sea floor as it must have been all those aeons ago. It was, in truth, a thrilling sight. The Leicestershire *Charnia* was perhaps a stray. Maybe it had become dislodged from its attachment to the sea floor and had been swept away. But here, in the rocks of Newfoundland, Charnias were in their true home. They were almost as thick as bluebells in a forest.

The time when they flourished was so early in the history of life that the difference between animals and plants had not yet become established. And their resemblance to fronds of bracken is only superficial. In fact they grew in a very different way – a way that resembles the one by which frost forms patterns on a window pane. It's called 'fractal'. One starts as a small blob. After a little time that puts out a branch with a particular shape. That branch in due course puts out two other similarly shaped branches. Each of them then puts out another two branches so that a structure grows which is based on one pattern repeated endlessly. It is one of the simplest ways of increasing in size for it requires only a very simple genetic mechanism. One of the scientists studying these fossils told me that he would only need to tap about eight commands into his computer to create a similar fractal pattern on his screen.

No animal alive today grows in such a simple way. It seems that the method does not allow enough variety to enable the organism that uses it to evolve such things as mouths or legs. And in the end, these fractal organisms were displaced by creatures with more complex ways of growing that were capable of producing more complicated structures.

Since those discoveries, the search for Precambrian life has spread and intensified. There have been finds in Siberia, the Ural Mountains and southern Africa. More kinds of animals have been found. But *Charnia* remains one of the very first living organisms of any size of which we have any knowledge, part at least of the answer to the question that so troubled Darwin – what were the animals that lived during the first stages of life on earth?

I asked one of the scientists working on these wonderful fossils in Australia if he knew why the animal he was studying was called *Charnia*. He confessed that he didn't. "Well," I said, "it's after a patch of countryside far away in the middle of England where I spent my boyhood" and I told him the story. But as I did so, I couldn't help wishing that I hadn't paid so much attention to the accepted geological wisdom of the time and that *I* had been the schoolboy who had found that key fossil in the Charnwood.

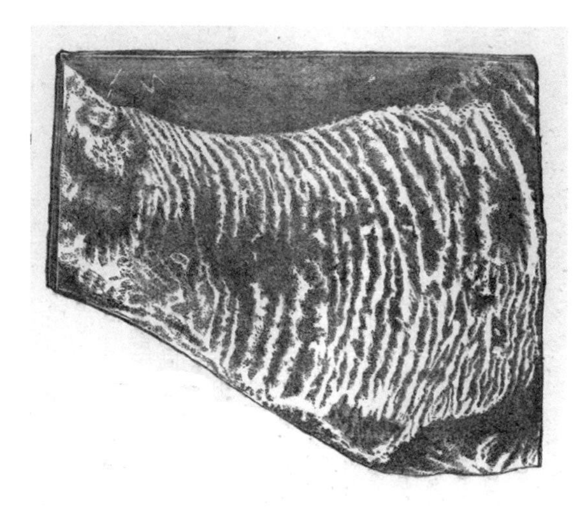

*This swirling pattern in some of Canada's most ancient rocks, was published by J.W. Dawson in 1875. His claim that it was evidence of early life was dismissed by many scientists, but is now recognised as being correct.*

Above. *Beacon Hill in Charnwood Forest is one of Leicestershire's most visited beauty spots, but it wasn't until 1957 that schoolboys climbing in a quarry nearby noticed the fossil that revolutionised our understanding of the first chapters in the history of life.*

Right. *One of the most perfectly preserved* Charnia-*type fossils to be discovered. It comes from Spaniards Bay in Newfoundland and is less than two centimetres long. The muddy deposit that entombed it, instead of squashing it flat, settled slowly around and within it, so preserving its structure in three dimensions.*

Above. *Fossils that are no more than faint impressions can best be examined by pressing a soft putty of some kind on to them and then holding it so that the light falls on it in the most revealing way. Jim Gehling, one of the foremost authorities on Ediacara fossils, shows how it is done.*

Below. *The latest insights into the Ediacara fauna have come from plotting the distribution of the fossils on the ancient sea floor. To do that, the researchers remove each layer of rock, piece by piece, and reassemble it nearby.*

Left. Dickinsonia, *from the Ediacara Hills of South Australia. Wrinkles on some specimens show that this creature was wafer-thin. This specimen is only two inches across but other closely related species have been found that are 81 centimetres (32 inches) in length.*

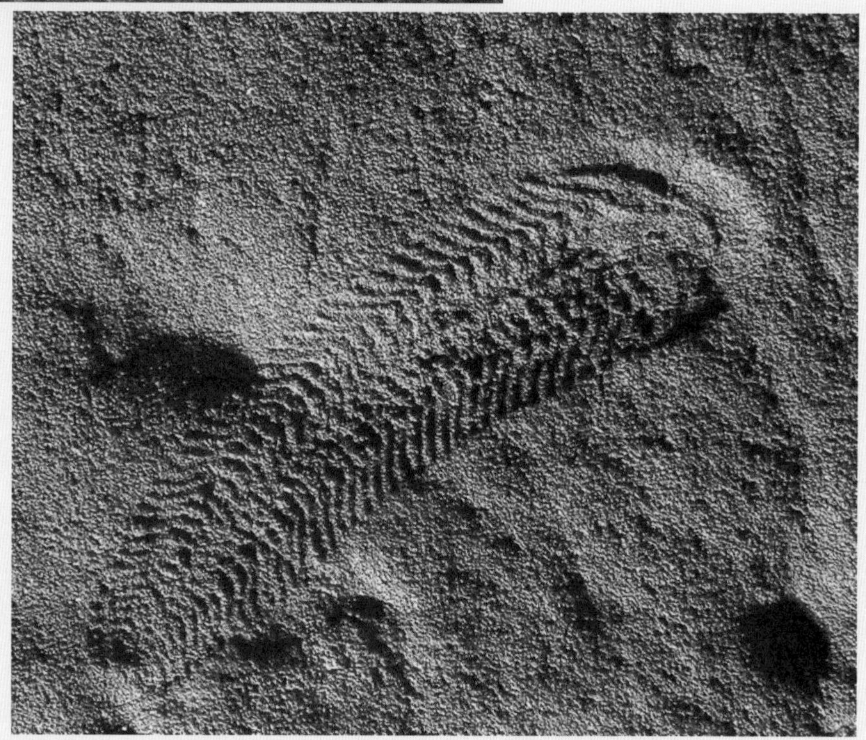

Right. Spriggina, also *from the Ediacara Hills. Although it is less than two inches long and the sandstone in which it lies is quite coarse-grained, its head-shield and separate segments are clearly visible.*

Left. *Guy Narbonne, who has spent most of his life studying the Newfoundland Precambrian fossils, holding a section of a frond of a Charnia species that was a metre in length.*

Below. *A fractal fossil,* Fractofusus, *from Mistaken Point. It grew by repeating over and over again a simple branching structure in rather the same way that ice crystals form a pattern on a window pane.*

Right. *Researchers remove a plastic mould from the surface of a 575 million year old Precambrian sea floor, at Mistaken Point. Back in the laboratory they will use it to work out the spatial relationships between the many hundreds of fossil impressions it carries.*

Right. *A group of Mistaken Point fossils. The discs may have been holdfasts with which the frond-like forms were attached to the sea floor.*

Fructus Artischochi.

# 4

# Foreign Fare

Why do you suppose that there are two kinds of artichoke – globe artichokes which look like large thistles and Jerusalem artichokes which look like potatoes? Come to that, why are there two kinds of potato – normal potatoes and sweet potatoes?

◁ *The head of an artichoke, pictured by Basilius Besler and published in his great catalogue of the plants in a garden near Eichstatt belonging to a German bishop.*

Why do you suppose that there are two kinds of artichoke – globe artichokes which look like large thistles and Jerusalem artichokes which look like potatoes? Come to that, why are there two kinds of potato – normal potatoes and sweet potatoes?

The fault, if fault it is, can be traced back to Columbus. One of his main purposes in sailing across the Atlantic in 1492 was to try and get to the source of those most valuable of domestic commodities of the time – spices such as ginger, cinnamon or nutmeg. Everyone needed them, if only to take away the terrible taste of the near-putrid meat that in those days before refrigeration was so often served up at mealtimes during the winter. Spices came from far away to the East, somewhere in the then mystic Orient, with everyone who handled them on the way adding their commission to the price. In consequence they were appallingly expensive.

Columbus' idea was that by sailing west across the Atlantic Ocean, he would get to Asia, the source of spices, by the back door, as it were and so cut out a legion of middlemen. So when he and the settlers who followed in his wake landed in the New World, one of the first things they did was to go around tasting the leaves and fruit of the local plants, all of which were entirely new to them, to see whether in fact they could be the source of spices. And if they discovered one the taste of which vaguely reminded them of one of those valuable commodities, they gave it the same name.

When this new kind of artichoke got back to Europe, gardeners who were perhaps a little more botanically minded, recognised that it was a kind of sunflower so they called them 'girasole' artichoke – 'girasole' meaning a plant that 'turns with the sun'. Eventually this new kind of artichoke reached Britain, and we, with our splendid insularity, did not deign to call them by a Spanish name. Clearly what these foreigners meant by 'girasole' was Jerusalem. And it was with that name that the new kind of artichoke appeared in our cookery books. And then, when eventually someone for some reason made soup from the Jerusalem artichoke, they thought it perfectly sensible to call it Palestine soup.

As for the potato, well that was slightly different. It was the sweet potato that got here first. People in Haiti, where it was first found, and where Columbus first landed, call these tubers 'batatas'. They came from a climbing plant related to our bindweed, convolvulus, and they were well worth eating. So they were sent back to Europe. In fact, they were almost certainly the first American plant to reach Europe alive. European cooks were quick to see their value, but this new vegetable didn't spread widely because coming from the West Indies it needed a relatively warm climate. So it couldn't be cultivated in the northern parts of Europe. But it did well enough in the south and soon became popular, partly perhaps because for some reason these potatoes were thought to be aphrodisiac – 'inciting to Venus ' as one herbal put it, or more plainly, 'procuring bodily lust.'

It wasn't until some time later – around 1570 – as Europeans fought their way bloodily across South America and settled permanently in these new lands, that they found, in colder regions high up in the Andes, a quite different plant but one with tubers that looked somewhat similar to the West Indian potato. So they called that 'patata' as well. This one, however, coming from a colder climate grew very well in the northern parts of Europe. And it was a huge success. It produced five times as much food per acre as any native European crop. Before long it became the more widespread and commonly consumed kind of potato. So it took that name, and its predecessor, the relative of the bindweed that had been the first potato to grow in Europe, became known as the 'sweet potato'.

The new potato from the Andes belonged to a huge plant family with around fifteen hundred species worldwide nearly all of which produce either berries or leaves with very remarkable properties. There are only five that are native to Britain. One of them, woody nightshade – *Solanum* in botanical Latin – gives its name to the whole family, the Solanaceae.

The strange properties of solanaceous plants have been known in Europe since medieval times. The leaves of woody nightshade were thought to be a good cure for rheumatism and also, as the 17$^{th}$ century herbalist Dr Culpepper put it 'excellent good to remove witchcrafts in both men and beasts as also all sudden diseases whatsoever.' The leaves of henbane, another member of the family, contain a sedative which, when concentrated, is lethal. It was the poison of choice for Dr Crippen.

And then we also have, perhaps most notorious of all our solanaceous plants, the mandrake. Every part of it – roots, leaves and fruit all contain a substance that if swallowed can produce hallucinations. Its root is like that of a small parsnip and is often divided into two so that with a little bit of imagination it can be interpreted as a human figure with its divided root being a pair of legs and its leaves a shock of hair. You might even – by inspecting one with great care – be able to decide whether it is a mandrake or a womandrake. Medieval herbalists maintained that digging up a mandrake plant was an extremely dangerous business because the plant screamed when you pulled it up and if you heard that, you would be struck dead. So, they claimed that in order to get the mandrake root that was so crucial for their potions, a man had to approach the plant in the middle of the night, with his ears plugged with beeswax, and taking with him a trumpet and a dog.

He had to tie the dog to the mandrake and at the stroke of midnight, whip the dog while simultaneously blowing the trumpet. So the dog bolted, the mandrake was up-rooted, the man heard nothing – and the herbalist got the substance he needed for his very expensive potions.

The solanaceous fruits found in the New World have, in sober truth, even stranger properties than the mandrake root. Apart from the potato, whose berries incidentally really are extremely poisonous, there is one species that produces bright scarlet berries. When the European settlers sank their teeth into them, their lips burnt. Here, no doubt they said to themselves, is what we really came for. This really *is* a spice and no doubt about it. This is a pepper.

So now once again, we have got two different plants with the same English name. The first peppers were small nuts produced by a vine that grows in the jungles of India; and then came New World peppers that are soft, fleshy and red. These new red ones have become so popular that they have been bred into several different forms. Not only do we have the little red ones that make you think your mouth is on fire, but bigger ones that can be either bright red or a luscious green and are rather less aggressive.

But there is one fruit from the Solanum family which, in popularity, rivals the tubers of the potato. In its original form its berries are yellow. When it first arrived in Europe, the Italians called it 'golden apple'. The French gave it their own name – 'love apple.' To my palate at least, it is the tastiest of all the products of solanaceous plants. It is not like any spice from the Orient, or indeed anything else.

It has a taste all its own and a smell, if you grow it in your greenhouse, which I find truly thrilling, a wonderful musty, tangy aroma that takes me right back to the jungles of middle America where it came from. And because it is in no way in the least like anything else in our kitchens, we have no other name to give it except that used by those who originally cultivated it. They were the Aztecs of Mexico and they called it 'tomatl'. That doesn't come easily to European lips, but we make a reasonable stab at it. We call it 'tomato'.

Above. *The first illustration of a potato plant, published by John Gerard in 1597 in his Herball. The plant came to Europe by way of North America and was known to him as the 'Virginian potato.'*

Above. *The sweet potato has long thin stems like its relative the bindweed, though it does not climb but lies flat on the ground. Its tubers are pink or purple and do have a sweet taste.*

Below. *The potato originated in the chilly highlands of the Andes. The people there still grow it as one of their most important crops just as they were doing long before any Europeans reached them.*

*The earliest surviving herbal was written by a Greek, Dioscorides, in the first century. It was copied and re-illustrated for many centuries thereafter. This picture of the mandrake comes from a manuscript copy of it dating from around 1458.*

Above *Although tomatoes are often called vegetables, they are in fact fruits. This wild kind, with black fruits, comes from Peru.*

Below. *Red peppers. Their burning taste comes from a substance in the fruit's internal walls. The outer walls are quite free from it and have a mild taste.*

# 5

# Cicada

In 1633, thirteen years after the Pilgrim Fathers had landed on Plymouth Rock in New England, their little colony was struck by a plague. Suddenly, and within the space of just four days, their houses, their fields, their orchards were covered by literally millions of red-eyed insects.

◁ *Cicadas like these, on a tree in Baltimore, only appear once in seven years. In between times the insects, as larvae, are hidden underground, attached to tree roots and drinking the sap.*

In 1633, thirteen years after the Pilgrim Fathers had landed on Plymouth Rock in New England, their little colony was struck by a plague. Suddenly, and within the space of just four days, their houses, their fields, their orchards were covered by literally millions of red-eyed insects. They were a good inch and a half long and had huge transparent wings which they held closed over their somewhat shorter bodies. These extraordinary creatures emerged from the ground and crawled up into the trees and onto the house walls. And there they sang. The noise was ear-splitting. Somewhat like a circular saw hitting a metal nail. People had to shout to make themselves heard. One farmer said he couldn't even hear the bells around his cattle's necks although he could see that they were ringing.

The Pilgrims, of course, were devout Christians and some of them maintained that this was a plague of locusts, like that with which Jehovah threatened the land of Egypt. According to the Bible, Jehovah said that 'they would cover the face of the earth that one cannot be able to see the earth and they should fill the houses and the houses of thy servants and the houses of all the Egyptians which thy father, nor thy father's father have seen since the day that they were put upon the earth unto this day.' That certainly fitted.

But other colonists disputed this. The Bible also says that these locusts, when they did come, 'did eat every herb of the land and all the fruit of the trees and there remained not any green thing in the trees or in the herbs of the field.' And these red-eyed bugs didn't do exactly that. They did, however, destroy much of the vegetation because they laid eggs on the leaves which then turned brown. Maybe they were not locusts but giant flies. Whatever they were, they were certainly a plague and unlike any that the settlers had seen before. And then suddenly, after some three weeks, all these strange insects died. Hundreds of thousands of the bodies lay in great drifts over the face of the land and were eaten up by the colony's pigs. The plague, whatever it was, was over.

Nor did it reappear the following year – or the year after that. But in 1650, seventeen years after the first plague, the red-eyed insects reappeared. Once again the land echoed with their calls and once again, within a month they all died. Another seventeen years passed – and it happened again. And so it has gone on happening at seventeen year intervals ever since. For these extraordinary insects were not flies. Nor were they locusts, though some local people still today call them that. They were cicadas – periodic cicadas.

The eggs, laid by the females in slits that they cut in stems and branches, hatch after six to eight weeks. The larva, a small maggot, drops to the ground and burrows into the soil. When eventually it encounters one of the tree's roots, it stabs it with its thin tubular mouthparts and starts to suck out the tree's sap. Most insect larvae as they increase in size go through several moults. In many cases that happens every few weeks. But the tree sap on which the cicadas live is so watery and lacking in nutriment that they grow very slowly and only moult every few *years*. Each time they do so, they detach themselves from the tree root and then, having shed their old skin and allowed their new one to harden, they burrow upwards a few inches before reattaching themselves to a root. They remain underground, for exactly seventeen years. Then, simultaneously, they all start on the last lap of their long journey and arrive at the surface.

But they don't all break through to the daylight immediately. Instead each builds a little mud turret around itself on the surface and there it waits. Perhaps that is to allow those that are a little late to catch up. Maybe it is to wait for the weather to be exactly to their liking. Whatever the reason, the moment comes when they all start to emerge together. As they crawl out, their skins split and reveal the fully-formed, red-eyed adults. In thousands. In tens of thousands.

How, lying in the darkness below ground, do they know when the time has come for them to return to the surface? It could be that they are aware of the passage of each year by the changes in the sap they are drinking which clearly is not the same in winter, when tree growth has virtually stopped, as it is in summer. But that only puts the puzzle back a stage. Even if they do know that once again summer has come, how did they count and know that this is the *seventeenth* summer. It certainly can't have anything to do with the weather in any one particular year because there are a dozen different populations of these cicadas in North America each of which keeps to its own seventeen year cycle – but starting in different years.

And just to complicate things still further, there is another different species that works on a thirteen year cycle. It lives further south in the United States. In places, the ranges of these two species overlap and someone has worked out that every two hundred and twenty one years, the lucky – or unfortunate – inhabitants of a few cities in the central United States can expect the simultaneous arrival of both species – a double dose, as it were. The fact is, however, that no one has yet worked out what clock any periodic cicadas use to coordinate their lives.

Nonetheless, they do all emerge within three days. Why should they all do so simultaneously? Well, it certainly ensures that males will find females. And it also is a good survival tactic, for the local birds and other creatures in the neighbourhood that might eat them, cannot possibly deal with the enormous numbers they are faced with. So the great majority of cicadas survive to breed – and once again the people are deafened.

It is only the males that call. They have a chamber on either side of the first segment of their abdomen, which has a circular membranous wall. This is attached to a muscle which, as it contracts, distorts the membrane so that when the muscle relaxes, the membrane snaps back. As it does so, it makes a click in much the same way as you can make a click with a dented tin-can by pressing it – except that the cicada can do that up to 180 times a second. The noise is then amplified by a huge, internal resonating chamber, so large that it occupies most of the insect's abdomen and its gut, stomach and other internal organs have to be squeezed into a small section at the far end.

The males call as they fly. The females sit on the branches listening. When one is ready to mate, she responds with a quick upwards flick of her wings. This makes a clicking noise – faint by comparison with the deafening calls of the males. You can't help wondering how the males can hear it in the din they are making. But you can easily prove that it is the sound of the wing flick that attracts the male because you can call him with a click of your fingers. I tried it. A male zoomed by and I snapped my fingers. He changed course, visibly. I clicked again and he landed on my wrist and I was able to persuade him to walk up my forearm as I snapped the fingers of my other hand in front of him.

We filmed all this in the grounds of a nunnery in Baltimore. It was a beautiful setting with plenty of small trees the branches of which were low enough for us to see that they were thick with insects. The director gave me my instructions. I had to start about fifteen yards away from camera, walk towards it, speaking as I came and end up immediately behind a horizontal branch that was at roughly eye level and covered with singing male cicadas. I set off.

'Stop, stop, stop!' said the cameraman 'Look behind you.' And there, some distance away was a small minibus slap in the middle of picture. As we watched, the door of the minibus opened and a dozen or so old ladies emerged. Each of them carried a long wand, holding it vertically in front of her and started to look for a place to stick it into the ground.

We walked over and explained what we were doing. They were most interested. Yes indeed, the locusts are extraordinary, aren't they. Most interesting. And what were *they* doing. 'Well,' one of them told us, 'We have spent our lives very happily in this nunnery and now we are selecting the place where each of us would like to be buried.' So as we were filming creatures in the foreground clambering up out of the earth, they in the background were making arrangements to go down into it. They thought the notion was highly entertaining. 'And will we be appearing on television?' they asked. But the director said that he thought it would be better if we re-shot the scene.

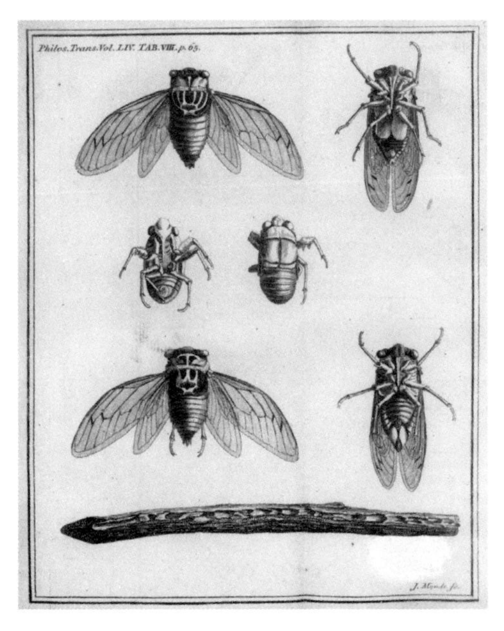


*The first English description of periodic cicadas was published in 1764 in a paper read before the Royal Society by Peter Collinson. He illustrated it with this engraving. Unfortunately, the insect represented is an ordinary American cicada that appears every year. Perhaps Collinson didn't know the difference.*

*Three stages in the emergence of an adult cicada from its larval skin. At first it is a ghostly white. After about an hour, its huge transparent wings begin to expand. Then it darkens and gradually acquires the jet black body and scarlet eyes of the adult.*

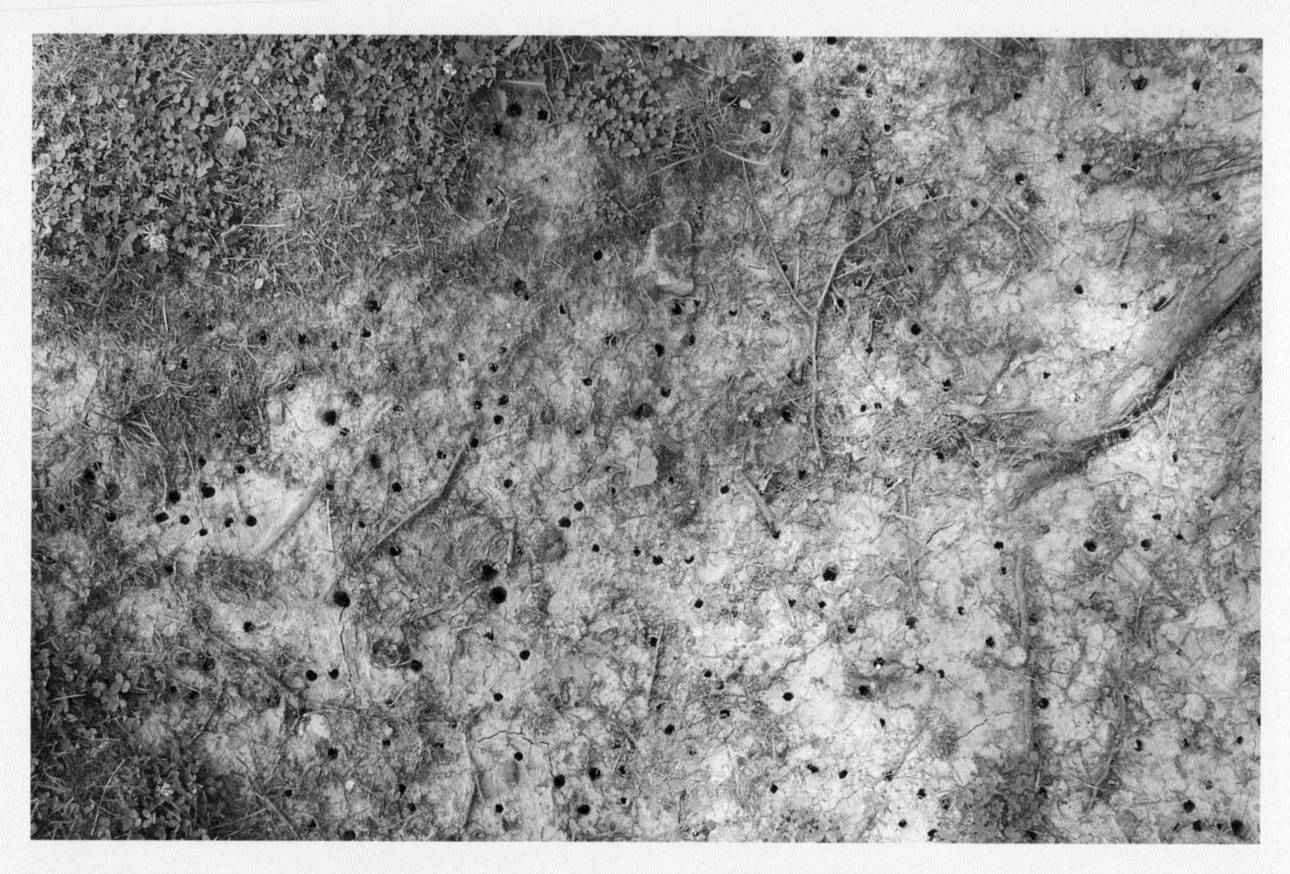

Above. *The ground beneath a cicadas' tree is riddled with holes from which the adult insects have emerged.*

Below. *The empty larval cases accumulate in great drifts.*

Above. *An adult male cicada with its wing lifted to expose the tymbal, the taut ribbed membrane which, when vibrated by a muscle, produces the insect's droning call.*

Right. *An adult male cicada. It is about an inch and a half long. It does not eat and after four to six weeks, it dies. Soon after the adults disappear, the larvae hatch from the eggs left behind by the female and dig themselves into the ground. There they attach themselves to tree roots and begin drinking sap.*

# 6

# Earthworms

I can't truthfully say that the meadows of Gippsland in southern Australia are exactly like the rich pastures of Somerset. The grass is considerably dryer, the trees in the spinneys are eucalypts and there are tree ferns growing in the gullies.

◁ *Giant earthworms occur in many parts of the tropics, including South Africa, Brazil and the Colombian Andes, but the best known is this species from southern Australia.*

I can't truthfully say that the meadows of Gippsland in southern Australia are exactly like the rich pastures of Somerset. The grass is considerably dryer, the trees in the spinneys are eucalypts and there are tree ferns growing in the gullies. Nonetheless, these grasslands are grazed by cows and sheep and as I wandered across them, I felt quite at home. Until, that is, I heard a most extraordinary noise behind me.

The noise, in itself, was not all that extraordinary. Indeed, it was the very fact that it was familiar that made it extraordinary, for it sounded very much as though someone just behind me had pulled a chain and flushed a lavatory. I turned round – and the field was totally empty. Not even a cow with digestive problems.

But I did know what had made the noise. Indeed, it was why I was there. It was the sound of a giant earthworm, squelching its way through one of its long tunnels, a few feet down in the water-logged soil. That noise, however, is about the only sign you can find of the presence of a giant earthworm unless you start digging or – as we did – you come across a cutting that was being made for a new road. There we saw holes in the soil five or six feet below the surface level that were an inch or so across. And in one of them, we eventually found our worm.

It was about as thick as my thumb but about six feet long. I say 'about' because measuring worms is somewhat problematic. A worm, after all, stretches. That in fact is the method it uses to move along its tunnel, pushing its front end forward through the soil, sticking out short bristles to hold it in position and then contracting the muscles lower down in its body and hauling up the rest of it.

Several people have been anxious to claim that they had found the biggest of all, the world-record holder. But you have to curb your enthusiasm when you measure giant earthworms. They are not only extendible, they are breakable. Even so, one was recorded as being thirteen feet long and managed to survive the measuring process in one piece.

There are lots of different kinds of earthworms. They are found on all continents – except Antarctica where the earth, such as there is of it, is frozen to a great depth and would be impossible for a worm to tunnel through. Continental Europe has around two hundred different species. We in Britain are not so well off. It seems that the freezing of our soil during the Ice Age exterminated all we had at that time. When the glaciers started to retreat, worms from the warmer un-glaciated parts of Europe began to move northwards but their progress was, as you might imagine, rather slow. And when the rise in sea level created by the melting glaciers caused an arm of the Atlantic to flood across the continent, so forming the North Sea and cutting us off from Europe, only twenty six species of its worms had managed to reach us.

They, however, have flourished here. There can be as many as fifty thousand in an acre. Even a modest little lawn is likely to contain around a hundredweight or so. And we should cherish them for they are invaluable in creating a fertile well-aerated soil.

The kind we see the most of, the one that comes out on to the surface of the grass at night, and the one that the early bird usually catches, is called technically, *Lumbricus terrestris*. Darwin studied it intensively and wrote a book a hundred and thirty years ago that has still not been superseded. He made a long series of simple but revealing experiments and eventually concluded that earthworms are intelligent.

At night they come out on the surface to nibble fallen leaves. They also drag them back to their tunnels to block up the entrances. It is not absolutely certain why they should want to do this. Darwin thought that they probably did so in order to prevent cold air drifting down their tunnels – the evidence being that when he kept them in pots in his warm study, they did so in what he described as a rather slovenly manner. However that may be, it was the way in which they transported their leaves that particularly interested him.

When they find a leaf, they inspect it and then grip it with their mouth and drag it back home. But they don't grip it just anywhere. Clearly it is easier to drag a leaf into a narrow burrow by its narrower end. To some extent, that depends on what kind of leaf it is. Pine needles, for example, they grip by the lower end. But most leaves they grip by the tip and not the stalk. They are able to distinguish between different kinds of leaves by their taste, but it was the shape that determined how a worm dealt with one. Darwin proved that by offering them differently shaped pieces of paper. They were all triangular but their overall shapes were ones that the worms had not encountered before. They crawled all over them, and then, nearly always, managed to select the sharpest corner as the one by which to pull it into the mouth of their tunnels. That requires some intelligence, particularly if you are totally blind.

Worms are also sociable creatures. When they emerge on the surface at night, they explore their surroundings, poking their heads into neighbours' tunnels, while holding on to their own with their tail. And on damp nights in the summer, they mate with one another.

Each individual worm is able to produce both eggs and sperm. Each has a broad yellow band encircling its body and extending across half a dozen of its segments. This is not, as some suppose, the place where a broken worm has joined its two bits together. It is a special area rich in slime glands that is crucial to their reproductive procedures.

The process starts when two neighbours that might have encountered one another on a number of previous occasions during their nightly explorations, meet and, instead of going their separate ways, linger and lie alongside one another. Their glandular bands begin to produce slime and in such quantities that it covers both their bodies and effectively seals them together. One of them then produces sperm from its testes which are placed in the tenth and eleventh segments counting backwards from the head. The sperm slowly travels along the tube formed by their two conjoined bodies until it reaches a special little pouch in the flank of its partner. That achieved, the other reciprocates by sending sperm in the opposite direction to its partner's pouch. The two bodies pulse slowly as this is happening and the whole process may take three hours or so. When at last it is over, the neighbours retreat to their respective tunnels.

A few days later, each worm produces more slime from that yellowish band. This time, however, the slime slowly hardens into a ring. The worm starts to jerk its body and the ring moves away from the glandular band that created it and starts to travel up the worm's body towards its head. After passing a segment or two, it arrives at the one that contains the pouch holding its neighbour's sperm. Out the sperm comes and the ring moves on taking the sperm with it. On it goes, along the worm's body, passing several more segments, until it arrives at the one in which it holds its own eggs. They too are collected and held beneath the ring where they are fertilised by the sperm. Then once again the ring moves on taking the now fertile eggs with it. Eventually the ring reaches the head. As it moves over it, the front part puckers together and seals itself – as it does at the back – and the worm moves away leaving behind a neat brown oval-shaped capsule lying in the soil.

That however, is how our garden worms do it. One assumes that the Gippsland giants have the same sort of technique. We did indeed find one of their egg capsules. It was rather more elongated than the ones I have come across in my garden – and of course, it was huge! The size of a cocktail sausage. I held it up to the light and could see the baby giant slowly writhing about within. But whether or not it produces its young using the same sort of assembly line technique as our own earthworms, I do not know. Indeed no one knows. For the Gippsland giant has never been recorded as coming to the surface. So copulation must go on in those subterranean tunnels. Perhaps there are special double width tunnels or sidings as it were where partners can lie alongside one another, though no one, as far as I know, has found such things. But I do sometimes speculate what sort of gurgling and squelching noise those subterranean embraces must produce in the meadows of Gippsland.

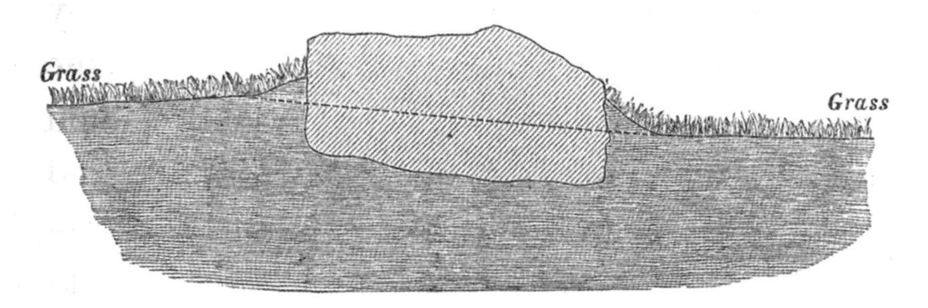


*Darwin concluded that the activity of earthworms was responsible for the fact that the fallen stones of Stonehenge such as this one have now sunk below ground level. He went on to demonstrate this process using a large stone disc in his own garden in Kent, the subsiding of which he carefully measured.*

Right. *Earthworms pull leaves into the entrances of their burrows though no one is sure why they do so. They grip them by their sharpest point to make the job easier and remarkably, are able to select that point even though they are sightless.*

Below. *Copulating earthworms. Their two bodies are sealed together by mucus so that sperm of each is able to travel down the channel.*

# 7

# Wallace

*'The Malay Archipelago: The Land of the Orang-Utan and the Bird of Paradise: A Narrative of Travel with Studies of Man and Nature. 1869'.* Not exactly a snappy, whizz-bang title like, say *Green Hell* that was favoured by later generations of travel writers.

◁ *Alfred Russel Wallace aged 39 in Singapore in 1862 on his journey back to Britain after eight years of arduous wanderings in the islands of Indonesia.*

'*The Malay Archipelago: The Land of the Orang-Utan and the Bird of Paradise: A Narrative of Travel with Studies of Man and Nature. 1869*'. Not exactly a snappy, whizz-bang title like, say *Green Hell* that was favoured by later generations of travel writers. Nonetheless, when as a boy, I came across the book, my eyes widened with wonder. It was in two volumes and illustrated with wood engravings. The frontispiece of the first shows a man, naked but for a turban and a loincloth, being savaged by a gigantic orang-utan. That of Volume Two, was if anything even more exciting. A fuzzy-haired hunter is crouching in a tree, hidden behind a small screen of leaves, aiming an arrow at a bird of paradise.

I was already wildly enthusiastic about birds of paradise. They seemed to me, even then, to be the most wonderfully beautiful birds in the world. And then I discovered that the author, Alfred Russel Wallace, had been for a short time a teacher in a school in the very city where I then lived – Leicester. So this was a book for me.

The Malay Archipelago was the name Wallace used for the string of islands stretching between mainland Malaysia and Australia that we today call Indonesia. Wallace had spent six years there. It was not however his first expedition. When he was teaching in Leicester in 1844, aged 21, he had met another man, three years younger than him, called Walter Bates. He, like Wallace, was a fanatical collector of beetles and the two of them scoured the Leicestershire countryside in search of them. They found literally hundreds of different species.

Why are there so many species of animals? The question was being much discussed at that time. Was each species immutable or could one give rise to another? And if it did, what was the mechanism that brought that about? Wallace and Bates discussed it endlessly. Both of them yearned to go to the jungles of South America where there would be beetles of a variety and beauty beyond their wildest dreams. Together they worked out a way to do so. They would pay for their journey by collecting natural history specimens which they would sell, through an agent, to wealthy collectors.

In 1848, the two of them landed at Belem at the mouth of the Amazon. They worked together for two years but then they parted. Bates remained on the Amazon itself. Wallace explored its main tributary, the Rio Negro.

After four years away from Britain, Wallace decided to return to London taking with him his collection. Three weeks out at sea on the return voyage, his ship caught fire, and sank. With it went several thousand carefully prepared specimens, huge numbers of natural history drawings and all his field notes – the results of four years of arduous and taxing labour. It must have been heart-breaking – but he was lucky to escape with his life.

Nothing daunted, when he got back to Britain he immediately started to plan another major collecting trip. This time, by himself, he would go to the islands of the Malay Archipelago with his main objective the collection of birds of paradise which then were still little known. And in 1854 he set off again. After working in Java and Borneo he at last arrived in the Aru Islands, south west of New Guinea. And there to his huge joy, he at last saw birds of paradise and watched them displaying their breathtakingly beautiful plumes. That was a special thrill. He was the first European to do so.

He returned to his base on one of the Spice Islands, Ternate, with a great number of specimens of birds, snakes, lizards, butterflies – and of course beetles. Throughout his journeys he had been troubled by malaria and now he had another bout. While lying semi-delirious with fever, pondering yet again on the great species question he suddenly thought of a mechanism that might cause one animal species to give rise to a different one. Now it seems simple and obvious. Since almost all animals produce more offspring than will survive to breed, and since all are slightly different, the ones most suited to the local environment will be the ones to survive and hand on their characteristics to the next generation. So over many generations a new species will appear. When he recovered from his fever, he sat down and over a couple of days wrote out his theory in more detail. But how could he get it published? The best way, he decided, was to send it to someone in Britain to ask for advice. He chose someone with whom he had regularly corresponded on natural history questions, a wealthy country gentleman living in Kent called Charles Darwin.

For Darwin, Wallace's letter was a bombshell. He himself had had exactly the same idea some sixteen years earlier. He had known that it would profoundly shock the more religious sections of Victorian society so he had repeatedly delayed publishing it while he collected more proof. But now – how could he publish this theory as his own? It would seem to many that he had stolen Wallace's idea. He wrote in anguish to one of his friends saying 'I would rather burn my whole book than that he – Wallace – or any other man should think that I had behaved in such a paltry spirit.'

Eventually two of his closest scientific friends with whom Darwin had discussed his idea over the years – Joseph Hooker, the Director of the Botanic Gardens at Kew and Charles Lyell, the professor of geology at King's College London – decided that the best way out would be for Darwin to write his own brief account of his theory and then arrange that both his paper and Wallace's should be read out, one after the other, at a meeting of the premier natural history society in the country, the Linnean.

It was done without Wallace's approval. How could they have got it in time? He was still out in the Malay Archipelago pursuing birds of paradise. When he eventually received an account of it, however, he seems to have accepted that the solution was perfectly fair. And in any case, his thoughts seem to have been on other things. He had now found another totally unknown bird of paradise, in the island of

Halmahera. It was quite unlike any other member of the family. Instead of plumes sprouting from its flanks, it had a pair of long white feathers dangling from the front of each wing that the bird erected when it displayed. And in his journal he wrote 'I consider it the greatest discovery I have yet made.'

By now he had been away from Britain for eight years, a lone European, living in conditions of the greatest hardship and regularly laid low by tropical diseases for which he had no real remedies. It was time for him to return.

When he got back to Britain he brought with him a huge collection of preserved specimens and two living birds of paradise. The birds he sold to the London Zoo. They gave him a hundred and fifty pounds for them – plus free entry to the Gardens. And he went to see Darwin. By this time, Darwin's great book *On the Origin of Species* had been on the market for three years. It had caused a huge outrage. Many Christians were appalled at the suggestion that human beings were related to apes. Scientists on the other hand were beginning to appreciate that this explanation of the mechanics of evolution was the most important and far reaching theory in the whole of biology.

The two men greeted one another with the greatest warmth. Wallace doubtless said what he had written to one of his friends. 'I do honestly believe that with however much patience I had worked and experimented on the subject, I could never have approached the completeness of Darwin's book, its vast accumulation of evidence; its overwhelming arguments and its admirable tone and spirit. I really feel thankful that it has not been left to me to give the theory to the world'. Darwin, for his part, wrote to Wallace saying 'You would, if you had my leisure, have done the work just as well and perhaps better than I have done it.'

The story of scientific discovery is full of examples of bitter feuds, life-long enmities, and accusations that one researcher has stolen another's results. So reading the correspondence between these two great men is a delight. It is true that both men independently had the same idea. But Darwin had gathered mountains of evidence and experimental results in support of it and that Wallace could not match. Wallace might have reasonably claimed that he had not been given sufficient credit – but he didn't. Instead he used the meticulous observations he had made on his long journeys to propose a theory about the distribution of animals that foreshadowed the discovery in the next century that continents drifted around the globe. And he wrote voluminously in support of the revolutionary theory about the way species originated which he was happy to call Darwinism. For me there is no more admirable character in the history of science.

And as well as all that, he wrote a thrilling – and, indeed, inspiring – travel book.

*One of the men from west New Guinea who, Wallace noted, took great pride in their 'mop-like heads of frizzy hair'.*

Above. *Men in the Aru Islands hunting greater birds of paradise as they dance in their display tree. The arrows the hunters use have heavy rounded tips so that the birds are stunned and their gorgeous and valuable plumes are not soiled with blood. Wallace was the first European to see a bird of paradise in display.*

Right. *Wallace spent most of his eight years in Indonesia living in locally built huts, like this one on the island of Waigeo in which he spent six weeks. He slept in the upper compartment. The ceiling of the lower space was so low that Wallace, over six feet tall, thought of removing it. But then he found that 'by bending double and carefully creeping in, I could sit on my chair with my head just clear of the ceiling.'*

Below. *Darwin's home, Down House, in the Kentish village of Downe. A wealthy man, he lived here with his family and servants for the last forty years of his life.*

Above. *The village of Dobo in the Aru Islands south of New Guinea. This was where traders, like the Chinese man on the left, came seeking bird of paradise plumes. Wallace himself, in his hat, stands in the middle of the picture. A tree kangaroo, doubtless a pet, sits in the foreground.*

Right. *In his old age, Wallace was greatly honoured. The Royal Geographical Society and the Linnean Society both gave him their Gold Medals and in 1908 he was awarded the Order of Merit, which he found 'quite astonishing and unintelligible'. He died two years later at the age of 90.*

*The standardwing bird of paradise in display. Wallace considered this species, which was named in his honour* Semioptera wallacii, *to be the greatest of all his discoveries. It lives on only three islands in the northern Moluccas, west of New Guinea. A male, when displaying as here, spreads an iridescent green cravat and erects two long-quilled narrow feathers, the 'standards' that, when the bird is at rest, hang from the front of each wing.*

LODDIGESIA MIRABILIS.

J. Gould and H.C. Richter, del. et lith. Walter & Cohn, Imp.

# 8

# Hummers

The copper-belted puffleg;
the sparkling violetear;
the fiery-tailed awlbill;
the hook-billed hermit;
the *white-whiskered* hermit –
what kind of creatures could
have names like these?

◁ *The marvellous spatuletail hummingbird as pictured by John Gould. The two males in flight appear to be displaying to a perched female. In fact, however, the female of this species, which had not been discovered when this plate was drawn, lacks the extraordinary spatule-tipped feathers.*

The copper-belted puffleg; the sparkling violetear; the fiery-tailed awlbill; the hook-billed hermit; the *white-whiskered* hermit – what kind of creatures could have names like these? well, the mention of 'bill' gives a clue. They are birds. And the name of another – the sapphire-spangled emerald – gives a further clue suggesting that they are jewel-like. They are indeed. They are hummingbirds.

Those names – and many more like them – come from one of the most magnificent of all bird books, six huge folio-sized volumes about the hummingbird family published by the great Victorian ornithologist, John Gould. It describes 320 different species and devotes a single magnificent hand-coloured plate to each and every one of them.

The Victorians, headed by Gould, were besotted by hummers, as their admirers call them. Nearly all of them are tiny. Indeed, among them is the smallest of all birds, the minuscule bee hummingbird from Cuba which is a mere two inches long – smaller than many a moth and much lighter than lots of beetles. Gould exhibited examples of all his three hundred and twenty species, carefully stuffed and mounted in twenty-four special glass cabinets during the Great Exhibition of 1851. He also produced for special customers, amazing assemblages of several hundred hummers all apparently visiting the same bush under a single glass dome.

There could scarcely be a less life-like display. The specimens are all male birds since in most species they are much brighter in colour than their comparatively drab females. Furthermore, a male hummingbird is usually extremely aggressive and will not tolerate other males anywhere near him. Nonetheless such trophies became a much admired feature in the drawing rooms of wealthy Victorians. The craze for them was such that during one week in 1881, the feathered skins of 400,000 of them were sold at auction in London to decorate the hats and clothes of fashionable ladies.

Put like that, it sounds appalling, but the spectacular beauty of the birds accounts for the mania. Their particular splendour is their iridescence. When the light catches their bibs or their crests, their moustaches or their pennants, they glint with all the brilliant metallic colours suggested by the names that Gould invented for them.

One of his most memorable plates shows a species he called the marvellous spatuletail. He gave it that name because the male has, projecting from his tail, a pair of naked quills, twice the length of his body and each tipped with a heart-shaped pennant. Gould, being Gould, shows not just one but two males in the air and another just beneath at rest. The two in the air however are apparently flying in the most amazing way with their pennant-tipped quills crossed perfectly symmetrically above their heads. When I first saw this plate, I thought Gould's artist must have used a considerable degree of artistic licence. No birds could possibly take up such a perfect posture in mid-air – and then a year or so ago I saw film of the male marvellous spatuletail performing his courtship display. He adopted exactly the posture shown in the picture as he hovered in front of the female. The effort involved in doing so was so great that rather like an Olympic weight-lifter, he could only maintain it for about five seconds. And then he had to drop away and have a rest on a branch to get his breath back.

The fact is that all hummingbirds are miraculously accomplished aeronauts. They can fly backwards; they can even fly upside down – and they can hang absolutely stationary in mid-air. To do that they have to beat their wings very rapidly indeed. One species can do so 75 times a second! Indeed their wing-beats are so fast that they make the humming noise that gives the birds their name.

They need to hover because they live almost entirely on nectar. Watch one feeding and you will see what consummate aeronauts they are. There's a buzz of wings and a tiny bird whizzes past you to a trumpet-shaped blossom hanging from a bush. There is no twig nearby, so no bird except a hummer can get near it. But the hummer hangs in front with its head only an inch or two away. Air, of course, is hardly ever completely still. There are always tiny currents even if only created by the beating wings of other hummers flying nearby and these inevitably blow the little bird about a bit. But the hovering hummer compensates for that by bending its neck so that while its body may move up and down a little, its head remains absolutely stationary with regard to the flower. And then with its wings beating at top speed, the hummer slowly moves forward through the air so that its needle-thin beak, with pinpoint accuracy, slides into the depths of the flower.

You can now see that the two halves of its beak are separating slightly. Its thread-like tongue is flashing in and out of the flower sipping its sweet nectar. After only a few seconds the bird has collected all the nectar available in that particular blossom and it moves backwards through the air, withdrawing its beak. There's another sudden hum and the bird has gone.

It is this nectar, rich in sugar, that gives them the energy needed to power their fantastic wing beat. Some species eat a few insects in addition, but all rely on nectar and the need for regular supplies of it dominates their lives. In some species a male takes possession of a single flowering bush and defends it against all comers. When it stops flowering the bird has to move on to another kind of plant with a different flowering season.

Heliconias, relatives of the banana, produce their flowers, one above the other, on a long dangling stem. But they only open one of them at a time and that only contains a limited amount of nectar. So a hummingbird that feeds on a heliconia has to fly off and find another plant nearby that has an open blossom still untapped, and then another one after that in a feeding circuit that is of just the right length so that when it gets back to the first heliconia, the plant has refilled its open flower with nectar And it fiercely defends its patch against any other hummingbird that might come and interfere with its finely tuned schedules.

Many hummers live in parts of the tropics where there are some plants in flower throughout the year. But not all hummers are so lucky. Some live in places where for a part of the year it is either too cold – or too hot – for plants to flower. Then the hummers must tackle longer journeys – *much* longer journeys. The rufous hummingbird for example – a tiny little creature 3¾ inches long and weighing only four grams spends part of the year in Mexico. But when summer comes, it sets off northwards heading for Alaska where it will breed. But it can't make that long journey on a single tank of fuel. It has to have regular refuelling stations all the way from Mexico through California, on into Oregon and up to Alaska. It gets there just in time to feed from the plants that for a few brief weeks will be in flower, before it has to head back south again.

The loss of a single feeding station along this two thousand mile long journey could be disastrous. Someone might chop down a crucial tree to make room for a house or a motorway so creating an uncrossable gap in the chain of refuelling stations. But people all along this route have come to welcome the arrival of the hummers. One man who delights in them, realising that his homestead is critical for the birds' success, put out an increasing number of feeders. More and more birds came to rely on him. He started by using a few kilos of sugar each season. Now he uses several tons. Other people, seeing what delight the hummers bring started to do the same thing. So what was a narrow migration corridor has now broadened and thousands of the little hummers are able to travel along a wider highway. And they are also able to spend even longer up in their summer quarters than they did in the past. Some are even appearing where none have ever done before.

We have become accustomed and saddened to read story after story of how animals have been stranded, decimated, and eventually exterminated by humanity's actions in destroying their natural homes. We rejoice when we hear of a last-ditch rescue mission that has managed to save one last sanctuary for a species that seemed doomed to disappear. But how satisfying it is – just for a change – to hear how human beings have not merely saved a species from extinction, but enabled it to spread and flourish better and more widely than ever before – particularly when the creatures concerned are as delightful and enchanting as hummers.

*By the beginning of the twentieth century, the fashion for wearing feathers was beginning to wane, helped by such mockery as this cartoon published in 1901 by the Westminster Gazette.*

Above. *Stuffed birds were highly fashionable decorations in Victorian drawing-rooms and hummingbirds were among their favourites. In most species, it is only the males that are spectacularly coloured so there is not a single female in this extraordinary assemblage of several hundred birds.*

Above. *A champion traveller, the rufous hummingbird. When autumn comes in the northern hemisphere, this tiny creature leaves its home in Alaska where it has nested and raised its young, and escapes the coming cold by flying south to Mexico. Next summer it will return – a round journey of about four thousand miles.*

Below. *The green violetear hummingbird lives in the cloud forest in the high Andes and not only collects nectar from flowers but uses its aerial virtuosity to catch flies and gnats.*

*Proof that the extraordinary mid-air posture of a displaying male marvellous spatuletail shown by Gould is virtually correct. The species is very rare and only found at very high altitude in the Peruvian Andes.*

*Fiery-throated hummingbirds in Costa Rica are not fussy about the source of their nectar. They collect it from trees, shrubs, vines and epiphytes like bromeliads. They also quickly learn to sip it from feeding bottles put out by human beings.*

# 9

# Identities

I have a blackbird in my garden – a male – who has a white feather in his left wing. I call him, rather unimaginatively, 'Whitey' and his arrival, a year ago, transformed my understanding of the dramas and battles that go on in my shrubs and on the bird table.

◁ *Blackbirds with white feathers are seen quite frequently in urban gardens. That may be because there are more people in towns to notice them. Alternatively, it may be because such conspicuous individuals are safer from predatory birds in gardens than in open countryside.*

I have a blackbird in my garden – a male – who has a white feather in his left wing. I call him, rather unimaginatively, 'Whitey' and his arrival, a year ago, transformed my understanding of the dramas and battles that go on in my shrubs and on the bird table. Suddenly I was aware how frequently – or infrequently – one individual bird visited my garden; how often he fed; whether he was likely to win an encounter with another male; whether he was courting; and what his relationships were with others of his own kind.

I'm sure that such a revelation comes to lots of us, but it is odd how long it took for naturalists to recognise the value of identifying individuals. Bird-lovers were among the first. Early in the nineteenth century the young John James Audubon, who would become one of the greatest of American ornithologists, tied silver threads around the legs of young flycatchers in their nest beside a creek near his house in Pennsylvania and so was able to prove that these small birds, although they disappeared southwards in the autumn, could nevertheless find their way back to the very spot where they hatched.

Even so, bird-ringing in a methodical and organised way did not start until almost a century later and those doing it were primarily concerned not with individuals but with the behaviour of a species as a whole. And that was true of people who studied not only birds – but animals of all kinds, even those most obviously individual animals, chimpanzees. One of the first to change this attitude was Jane Goodall, back in the 1960's. I suppose people must have known before then that individual chimpanzees could be recognised by their faces. But Jane took things further. She showed that individual chimps had sharply marked and widely varying characters. She irritated stick-in-the-mud scientists by giving her individual animals human names – Mike and Fifi, Melissa, Willie and David. That was like red rag to a bull as far as some zoologists were concerned. "Anthropomorphism!' they cried, accusing her of just about as big a crime as you could commit in zoology short of outright fraud – that of attributing human characteristics to an animal without incontrovertible evidence that they had them. Jane's meticulous studies, of course, eventually sorted that out and she was able to justify all her observations and deductions.

Individual chimps are comparatively easy to recognise – their faces are almost as varied as ours. But how do you identify other kinds of animals – individually? Slowly, as zoology moved out of the laboratory and into the field, scientists found ways of doing so. When a humpback whale dives, its large broad crescent-shaped tail often breaks the surface and waves in the air for a few seconds. Its underside carries black markings on a white background. Snap that with a camera and you have a record that will enable you to recognise that whale wherever it appears – and without interfering with it in any way. Such simple observations have led to extraordinary discoveries. Grey whales, for example, travel vast distances through the ocean every year. A female having given birth to her young in a warm lagoon off the coast of California, will travel northwards to spend the winter in the seas off Alaska feeding on the plankton that blooms there each year.

Elephants too are easy to recognise once you get your eye in. Their large flapping ears often get torn and become ragged, each in its own way. Echo, one of the most famous of Kenya's elephants who was studied continuously by Cynthia Moss for over twenty years until her death, was immediately recognisable even by those who had only seen her on television by the fact that her two long tusks crossed at the tip. And those who knew her, recognised that not only was she the accepted leader of her large family, but that she was exceptionally wise in her decisions.

Craig Packer, who studies lions, had a rather more difficult problem. How can you distinguish one adult lioness in a pride from another. Craig found the answer. Whiskers! It turns out that every lion has its own particular number and distribution of whiskers on its muzzle which remains the same throughout its adult life.

One of the most remarkable of these individual recognition systems, to my mind, was that devised by Peter Scott. He founded the Wildfowl Trust back in 1946. Some of the birds there are pinioned permanent residents, but they are joined each year by great numbers of migrants. Among these are Bewick's swans. They are somewhat smaller than our resident swan, the Mute swan and they have a lovely bugling call. But they also have another characteristic. Their beak is black at the front and yellow at the back. Where these two colours meet, they form an irregular line that may be zig-zag, curving or branching, and, what is more, each in a different way. Peter noted that and drew the beak of every one of his visiting Bewick's. More and more arrived in Slimbridge, year after year.

Within seven years of his first noting them, there were over a thousand arriving annually from the Arctic. The bill of each was drawn by Peter and his helpers, in particular by his daughter Dafila who, like her father is an artist as well as a scientist. Each was given its own name. That of course, risked the old charge of anthropomorphism but Peter maintained that names were easier to remember than numbers. So there were Suki and Kate, Liz and Cindy. And since Peter also had an irrepressible delight in puns there were also pairs, Pote and Tate, Hyla and – wait for it – Fling.

So he discovered all kinds of details about a Bewick's swan's life. Of the first thousand studied, there was not a single pair that split up to look for new mates. Youngsters fly down from the Arctic with their parents when they are four months old and so learn the way. While some become independent after the first year, others – less confident perhaps – stay with their family for up to four years.

It's not every species of animal that has such individual characteristics, at least to our eyes. In some cases of course they may have labels of identity that are beyond our senses. Our hearing is not as discriminating as that of a penguin. To us it seems inconceivable that the harsh croaks of a young penguin can be recognised among the continuous shrieking of a thousand other chicks in a colony. Yet we know that is so, for a returning female finds her offspring immediately even though it may have strayed far from the place where she left it. And we all know that a dog's nose is far more sensitive than ours since it can recognise all the individuals in a neighbourhood from sniffing their urine and immediately detects the arrival of a strange dog in the neighbourhood.

So this capacity to recognise individuals may be even more widespread in the animal kingdom than we realise. Does it apply, for example, to starlings? Does every one in a flock of ten thousand, wheeling across the autumn skies before settling down in their roost, have its own identity? Or every herring in a shoal of millions? It's hard to believe. But it certainly could be so.

I once made a film about spiders. One species I wanted to film was the bolas spider. It is common in parts of the United States and the females have a highly ingenious way of catching the moths that are their prey. Each hangs beneath a leaf and spins a thread of silk which has a blob at the end. She holds this with one of her forelegs. When she hears the sound of an approaching moth, she uses her leg to whirl the thread, in much the same way as a South American gaucho whirls his bolas. And then when a moth comes near, it gets entangled. And she hauls it in and eats it.

Kevin Flay had the job of filming them and went ahead to set things up. When I joined him, he had half a dozen of these spiders, each settled on a leafy twig, stuck in a milk bottle. Kevin took me down to see them. 'This one,' he said, 'is terrified of light. She hunts enthusiastically enough in the dark but does absolutely nothing when I turn the lights on. This one doesn't mind the lights but hates noise. And this one is just bone idle.' 'But this one,' he said affectionately, 'hunts no matter how much light I put on her or how much noise people make. She's a darling.' And so she did – and was.

So you see, animals of the same species are not necessarily machines. Even spiders have their own individual characters.

*John James Audubon, the great American ornithologist. He was one of the first to understand the importance of recognising individual birds, which he did by attaching silver threads to their legs.*

*Lions can be identified by the distribution of the bristles on their muzzles. That is made easier by noting odd isolated ones outside the long rows, as here.*

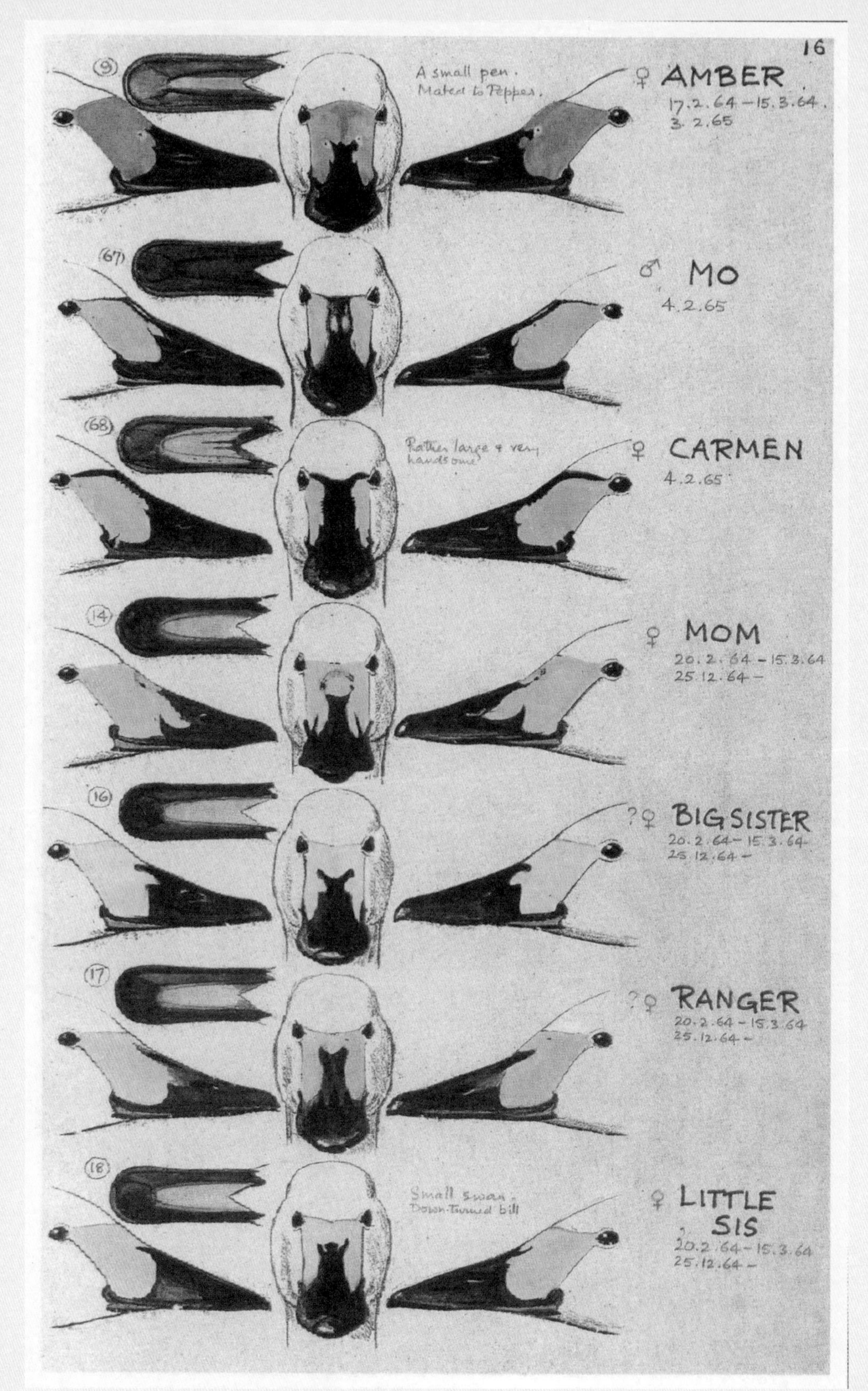


*In 1946 Peter Scott, with his artist's eye for detail, noticed that the Bewick's swans that came down from the Arctic each autumn to spend the winter at his Wildfowl Trust in Gloucestershire, could be identified by the black and yellow pattern on their beaks. This is a page from the catalogue he compiled that within a few years enabled him to identify over a thousand named individuals.*

Above. *Humpback whales always thrust their tails above the surface when they start their dives. The underside of each is marked with a unique black and white pattern. So a researcher having photographed the tail in Hawaii can send pictures to colleagues working in the Arctic who can then tell if and when that individual whale arrives with them.*

Below. *Bolas spiders may be difficult to identify individually, but their behaviour suggests that each has its own character.*

*It is not only human researchers who need to identify individual animals. A king penguin parent returning with a crop-full of fish for its furry-coated chick, needs to be able to identify it among a throng of several hundred. That cannot be done by remembering the place where it left it, for the chicks, in their crèche, are constantly on the move. Nor can it be done visually for all the chicks, to our eyes at least, look similar. Parent birds do it, almost unbelievably, by recognising the unique sound of their chick's braying call.*

# 10

# Rats

I have to admit that I hate rats. It's not a thing to be proud of. One should feel warmly about all of nature's creatures. But, nonetheless, I hate rats.

◁ *Brown rats are, surprisingly perhaps, relatively rare in the countryside. They seem to prefer to stay close to the houses and cultivations of human beings where so much food is so easily found. They are primarily burrowers, but also very capable swimmers, so the sewers built by man suit them very well.*

I have to admit that I hate rats. It's not a thing to be proud of. One should feel warmly about all of nature's creatures. But, nonetheless, I hate rats. I don't mean that I mildly dislike them as I dislike, let us say, maggots. I mean that if a rat appears in a room, I have to work hard to prevent myself from jumping on the nearest table. It is said, I don't know on what authority, that wherever you are in the settled world, you are likely to be within a few yards of a rat. There are rats in the rafters, within walls, and in the spaces beneath the floorboards. Most pervasively of all, they are in the drains and sewers that form a network just beneath our feet.

When I was filming in the Solomon Islands, we stayed in a village built on an artificial island in the centre of a coral lagoon. Paradise. You might think that you would be beyond the reach of rats there. But no.

The villagers gave us a hut in which to stay. It was thatched with coconut leaves and its walls were covered with pandanus matting tied to a framework of poles. As we went in and put down our gear, I noticed along each of the horizontal poles and about an inch above them, a dark greasy line. I knew what that meant. Rats. They made it as they brushed past the matting with their fur. Those poles were rat runways. So I put down my sleeping bag not against the wall as I might otherwise have done but right in the middle of the matting-covered floor.

That evening, just after we lay down to sleep there was a tremendous storm. Rain pelted the hut and the ground with a force that you seldom see outside the tropics. The noise of it beating on the trees outside was deafening. So at least, I said to myself, I wouldn't be able to hear the rats as they scampered along their runways. But as I lay with my eyes closed, trying to sleep, I felt a movement on the sheet around my feet. I flicked on my torch and there was a rat running across me. I looked around. There were rats everywhere. Henry the cameraman who had chosen to sleep by the wall, turned on his torch, saw them and said something unrepeatable. 'I'm not sleeping here', he added.

I agreed with him. The sound recordist and the director, as far as we could see, were sleeping through all this, so Henry and I thought it better not to disturb them. Clutching our sheets over our head we ventured out into the drenching rain to look for something better. And we found it. In the middle of the village we discovered a large building with a thatched roof supported by substantial wooden pillars but no walls. The floor was of neatly swept white coral sand and there were rows of benches. But no one was in it. And no sign of rats as far as I could see. So Henry and I lay down between the benches and at last got some sleep.

When I opened my eyes in the morning, I saw a ring of black faces, each surrounded by a white surplice, peering down at me. For a moment I thought I was in heaven. But no. We had taken refuge in the village church and the choir was assembling for morning service. The kind villagers thought that we were being absurdly over-sensitive about the rats – and perhaps we were – but they allowed us to go on sleeping in their church for the rest of our stay – provided we got up reasonably early.

A few years later I started work on a television series surveying the mammals of the world. We claimed that it would be fairly comprehensive so I could hardly leave out rats. There are over five hundred different kinds of them – alas! It would be cowardly to ignore such a successful and widespread species. So I decided to feature the commonest, *Rattus norvegicus*, the brown rat.

I should add here, in fairness to the Norwegians, that they should not be held responsible for this animal. The species apparently originated in southeast Asia south of the Himalayas. It reached Britain, on board Russian ships in either 1728 or 29, and didn't get to the Norwegian mainland until 1762. Now, of course, the species has spread right around the globe. However, if we were to film *Rattus norvegicus* at all, then I thought we should go the whole hog and show it in its most dramatic assemblies. And that, I reckoned, would be in one of the temples in India where they worship the brown rat.

I travelled to India with Trevor who was the sound recordist. Huw, the producer and Gavin the cameraman had gone ahead to make the arrangements and met us in a guesthouse not far from the temple. Everything was fine, Huw said. We had permission from the priest to film anywhere we liked in the temple and there were rats there literally in thousands. They were running all over the floor, he said enthusiastically – wherever you went. It was difficult not to tread on them. I smiled, a little wanly no doubt. 'Well, I shall make sure that I wear heavy boots and tuck my trousers into my socks,' I said bravely. 'No, you won't', said Huw, 'the temple is a sacred place. You aren't allowed to wear shoes. It's bare feet for you. But don't worry. I'll sit you on a high stool where they won't be able to get at you.'

I was not feeling well. For the first time in decades, I had a stomach upset. So after an Indian meal which I am sure was delicious but which I didn't consume much of, I retired early to the little lodge in the grounds where I was to sleep and sat down wearily on the lavatory. And a rat leapt up from between my thighs, scampered off and hid under my bed. I called, perhaps rather too stridently, for help. Gavin and Huw came and between them chased the rat from under my bed and then into the garden. There it disappeared down a drain, no doubt to be ready to greet me from the lavatory in the morning.

After breakfast the next day, Huw said he would go ahead to the temple and set up the camera and lights. Trevor and I could follow an hour or so later. The temple was easy to find. There were carvings of rats as decorative features all along the tops of its walls. A man sat at the doorway to prevent any dogs or cats getting inside and making an easy meal of the holy rats. Beyond him there was an open court, full of rats. Netting stretched above from wall to wall to protect them from hawks. And the rats swarmed in incalculable numbers, running in long processions along the floor close to the walls. A holy man sat cross-legged eating handfuls of rice from a large copper tray while rats ran over his shoulders and up over his matted hair. They even took rice from his fingers.

I could see filming lights shining in the inner court. That must be where we were expected. Treading gingerly between the scurrying rats, I made my way towards them. Huw was standing surrounded by rats beside a tall stool on which I was to sit, safely above the hordes. The man was clearly lacking in finer feelings. But to my astonishment, I saw that he was smearing the stool's legs with over-ripe banana. He was deliberately laying a trail to tempt the rats up the stool's legs and on to mine. 'It will make the sequence more dramatic' he said. But he did have the grace to look a little shame-faced as he said it.

I have some justification for being appalled by rats. For one thing they carry some fairly horrible diseases. They do, after all, live in sewers. But I suspect that my irrational horror of them comes from the fact that they live at such close quarters with us and while they sensibly keep out of our way when they can, they don't have any real fear of us. But looking at the situation as objectively as I can, I ought really to admire the brown rat. Some say that if the human race were to vanish, the rats would take over the planet. They can after all, eat and flourish on pretty well anything, meat as well as grain, rotten food as well as fresh. They can tolerate a wide variety of environments – heat as well as cold, light as well as dark. Female rats are very solicitous mothers. And they are highly intelligent with long memories. It is that combination of characteristics, in fact, that has led to their unparalleled success.

So, from one point of view, you might regard them as our greatest rival, both in numbers of individuals and in geographical distribution. Maybe if rats could talk, they might well say, with the same degree of distaste as we do of them, that wherever you go in vast parts of the habitable world, you are never more than a few yards from a human being.

*The temple sacred to rats in Bikaner, India, not only contains many thousands of living ones scampering throughout its courtyards, but also has marble ones running along the tops of the walls.*

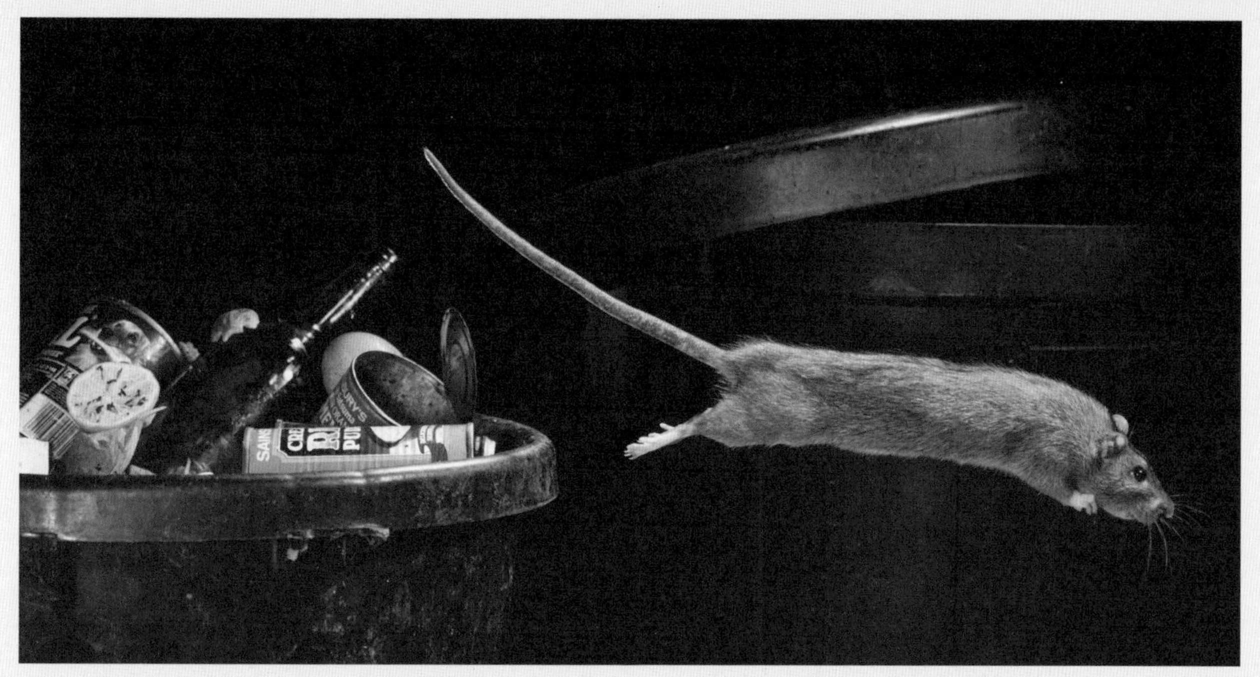

Above. *Brown rats, although they are not regular climbers, are nonetheless very agile.*

Below. *In and around human habitations, rats are protected from the privations of winter and can gather food the year round. They can start breeding when they are only three months old and produce five litters a year, with as many as twelve young in a litter. It is hardly surprising that they can suddenly infest a building.*

*The black rat is smaller and slimmer than the brown and prefers to climb rather than burrow. Once widespread in Britain, it was displaced by the brown rat in the eighteenth century and now there are only a few isolated colonies in Britain, mostly in seaports.*

*In India, there are temples sacred to rats. There, dogs, cats, hawks and anything else that might catch a rat are kept out and devotees happily allow them to share their meals of rice.*

# 11

# Monsters

Monsters have largely disappeared these days. Dragons, which were catalogued with such confidence and in such detail in the encyclopaedias of the sixteenth century, had gone by the end of the seventeenth.

◁ *This drawing of a sea monster attacking a ship is clearly based on a giant squid. A French naturalist, Denys de Montfort, published it in 1802. The animal was encountered, he wrote, by sailors off the coast of Angola in West Africa. They managed to prevent their ship being capsized by cutting off the monster's tentacles with axes and by praying to St. Thomas. On their return to their home port of St Malo, they immediately gave thanks in the saint's chapel and deposited there a votive tablet carrying a picture of the event. It was on this that de Montfort based his illustration.*

Monsters have largely disappeared these days. Dragons, which were catalogued with such confidence and in such detail in the encyclopaedias of the sixteenth century, had gone by the end of the seventeenth. Mermaids, peeking coyly from the waves to drive sailors mad with frustration no longer appear on maps drawn in the eighteenth. And the unicorn, nestling its spectacular head in the laps of maidens, has also – alas – been dismissed as a fantasy based on nothing more romantic than the overgrown molar of the narwhal.

But sea monsters? Well, they survived in our beliefs much longer – and not surprisingly. No one really had much idea of the world beneath the waves until Captain Cousteau developed the aqualung. That was less than seventy years ago and even now our knowledge of the really deep abyssal depths of the ocean is very spotty. So it is understandable that sailors once thought that there might be great monsters lurking in the depths. There were, after all, every now and then, clues. A huge tentacle writhing above the surface of the sea, we now know, is not an impossibility. A giant squid was once caught in the Bahamas that was 47 feet long. So the vivid drawings on seventeenth century maps showing ships entangled in them could well have some foundation.

Stories of a monster in the black waters of Loch Ness go back into Celtic antiquity, according to some Scottish scholars. And certainly, if you are going to have a monster in a lake, Loch Ness is the place to have it. It is the largest lake in the British Isles – 750 feet deep in some places – and it occupies a huge fracture in the earth's surface, the Great Highland Fault, that almost cuts the Highlands of Scotland in two. So it is not unreasonable to suppose that at some time back in the geological past – the Jurassic perhaps when there certainly were giant reptiles like ichthyosaurs and plesiosaurs swimming in the oceans – one or two of them living in the North Atlantic, made their way into the Loch and that their descendants have lived there ever since, glimpsed only by the occasional Highlander on the moors above or perhaps an early Christian hermit sitting beside his cell.

But then, in the 1930's a road was built running all along the north shore of the Loch and lots of people regularly travelled along it, looking over the bleak dark waters. In 1932, one of them reported seeing something strange, a sinister black shape rearing above the waves. Someone else produced a blurry photograph of such a thing which was published and monster hunters, to the delight and profit of those proprietors of restaurants and hotels along the road, flocked there armed with cameras with telephoto lenses and powerful binoculars and set up watch. Between them they produced a number of quite sensational accounts of sightings. One even reported seeing something writhing on the loch shore. People equipped with searchlights, sonar and powerful telescopes maintained a continuous surveillance of one particular bay for weeks on end. And then, in 1970, an American, Dr Robert Rines came with a team and the latest underwater cameras and sonar gear. And in 1972 he produced two pictures. They were taken at a depth of about forty five feet with a flash camera and seemed to show an elongated object, which had on its flank a diamond-shaped fin. It was a very long way from being conclusive but nonetheless the possibility of a monster existing in the Loch now seemed real and the search intensified.

Then, in 1975, I had a telephone call from Sir Peter Scott. He was at the time the leading figure in the conservation movement, not only in this country but worldwide. Would I meet him in the Natural History Museum as he had something to show me. He wouldn't say more.

So, of course, I went. We sat down at a table and he took something out of a small portfolio that he had brought with him and laid it on the table. "What do you think?' he said. It was one of his paintings and it showed two long-necked, small-headed fat-bodied creatures with diamond-shaped paddle-like limbs, their bodies dappled by light filtering down from the rippling lake surface above. They were not unlike plesiosaurs.

I took out my spectacles to have a closer look. I had only just started wearing them and, as it happened only that very morning I had been having some trouble with the frames and they had just been tightened by a cameraman friend. As I put them on, both lenses, simultaneously, popped out and landed with a tinkle on the painting. Peter was rather more startled than I was and for a moment he looked upset, even offended. He may even have thought that I was mocking him.

And he had every reason to feel sensitive, for the fact was that the Loch Ness monster had by this time become something of a music-hall joke. Responsible serious scientists poo-poohed the idea that it existed. Those in the very Museum where we were sitting had poured scorn on anyone who took the monster stories seriously. So Peter, by painting a pair as real creatures, was certainly going out on a limb. In fact he was being admirably brave. And he had a good reason. If there *was* a monster, and if someone got the chance of catching it or, worse, harpooning it, there was nothing to stop them from doing so. So in Peter's view, as a conservationist concerned about the loss of rare species, it should be legally protected. To do that it had to be given a name and something material was needed to demonstrate its existence A painting would do that – even if it did not prove it. So Peter had painted the monster's portrait and he was going to propose that it be given a proper scientific name. That, he told me, should be *Nessiterras*, after the Loch and *rhombopteryx* after the diamond-shaped paddles that had appeared in Dr Rines' photographs and now Peter's painting. I thought it was rather exciting – and I certainly admired his courage.

But the scientific world would not accept the existence of any species without some solid evidence – a bone, a scale, a bit of skin. And Peter's painting didn't qualify. Then someone pointed out that *Nessiterras rhombopteryx* was an anagram of the phrase 'monster hoax by Sir Peter S.' Two could play at that game and someone on Peter's side retorted that though the name was indeed an anagram, the phrase it was based on was actually 'Yes both pix are monsters'.

Now, sadly, much improved underwater detector gear has demonstrated beyond doubt that no great monster lurks in the Loch. And people in their old age have confessed that some of the photographs taken from the shores of the Loch were in fact deliberate hoaxes.

So is that the last of all the monsters? I cherish the hope that it is not. There is one possibility left that I know of. In 1951 the great British mountaineer, Eric Shipton, exploring the Himalayas not far from Everest, came across a line of human-like but immense footprints running across a snowfield. He and his Sherpas followed them for a mile or so before they disappeared onto rocky ground. There could be no doubt that they were real. Shipton photographed them. He was not a practical joker nor was he naive. He was a very experienced Himalayan traveller. Something – and something big – had made that trail. Whatever it was, it was clearly not a regular visitor to these high snowfields, for such prints were, if not unprecedented, extremely rare so presumably the creature must have wandered up from the thick rhododendron forest farther down the valley. The Sherpas, of course, had no doubt about what it was. It was the Abominable Snowman, the yeti, a kind of giant ape.

I still think that such a thing is not an impossibility. Back in the 1930's, a German anthropologist, Ralph von Koenigswald was wandering in the murky back streets of Hong Kong and idly looking through the extraordinary jumble of strange objects that are the stock-in-trade of a traditional Chinese druggist. And among the shrivelled snakes, bird's claws and fossilised bones, he found something that, so he said, made his hair stand on end. It was a molar tooth, just like a human molar, except that it was six times as big. There could be no doubt that it had come from a giant ape. After a long search in other similar places, he found two others. He published a description of them and called the creature they must have belonged to *Gigantopithecus*. We now know that they and other skeletal remains come from geologically recent deposits somewhere in Asia and that they are no more than about half a million years old. Their owners must have been alive just yesterday, compared to the theoretical Jurassic ichthyosaurs of Loch Ness.

So I am clinging to my hope that there really is a giant ape, an Abominable Snowman, somewhere in the remote valleys of the Himalayas; and that monsters haven't yet vanished from the face of the earth.

*In addition to squid-like sea monsters, there were thought to be others in the ocean like giant snakes that were capable of grabbing sailors from the decks of their ships. This picture appeared in a book by a Swedish Archbishop, Olaus Magnus, which he published in 1555.*

Above. *Fossils of gigantic marine reptiles dating from the Jurassic Period were first discovered and recognised in the 19th century. Philip Gosse, a well-known naturalist of the time, considered it quite possible that those known as plesiosaurs, which had particularly long necks, might have surviving descendants and these might be the origin of stories of monstrous sea-serpents. He published this picture to substantiate his idea.*

Below. *In 1951, Lachlan Stuart, who lived near Loch Ness, published this photograph of what he claimed was the monster. There were serious debates as to what it might be. It was not until some thirty years later that he confessed that the humps in the water were fakes he had fabricated.*

Right. *The mysterious flash photograph taken in 1972 by Dr Robert Rines and his team at depth of forty five feet in Loch Ness. Could it be the flipper of one of the reptiles that Philip Gosse had suggested over a century earlier?*

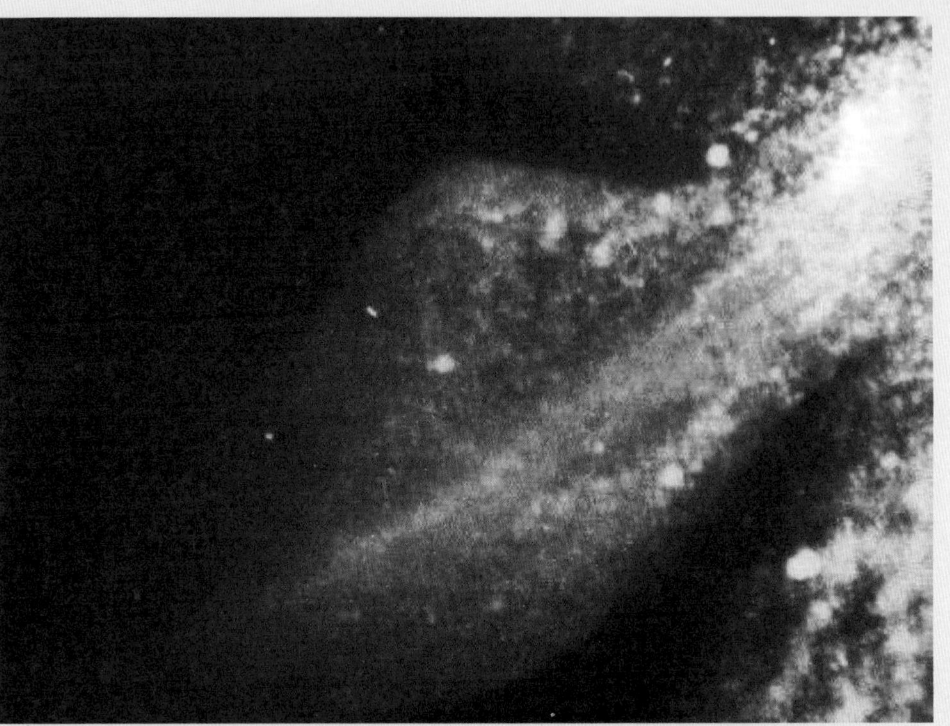

Below. *Peter Scott's painting which he hoped might help to establish the existence of an animal living in Loch Ness, to which the name* Nessiterras rhombopteryx *might be given.*

Could these be the footprints of the Abominable Snowman? This photograph was taken by the distinguished British mountaineer, Eric Shipton, in the Himalayas in 1951.

# 12

# Butterflies

A butterfly is, surely, the very image of delicacy and fragility. It seems almost unbelievable that it could be capable of doing anything more than flutter feebly around a flower. But that would be a very long way from the truth.

◁ *Monarch butterflies, having escaped the northern winter, assemble in millions in Mexico. They move as little as possible to save the tiny food reserves in their bodies which they will need to fuel their return journeys to North America.*

A butterfly is, surely, the very image of delicacy and fragility. It seems almost unbelievable that it could be capable of doing anything more than flutter feebly around a flower. But that would be a very long way from the truth.

I once stayed for a week or so in a tiny woodcutter's hut in the middle of a small clearing on the banks of a river in northern Paraguay. The rain-forest stretched unbroken for many miles all around. I was looking for armadillos, hummingbirds and boa constrictors. One morning, I swung my legs out of my hammock, peered through the door of the hut and was astounded to see that the clearing beyond was filled with a blizzard of butterflies. I couldn't even begin to make an estimate of how many there were. Certainly hundreds of thousands. Perhaps millions. The biggest were swallowtails. Some were velvety black with stylish carmine blotches on their hind-wings, others black and yellow, like bigger versions of the European swallowtail. There were also much smaller ones that had iridescent blue fore-wings and scarlet hind-wings with white undersides.

Down by the river, on the little muddy landing where we kept our canoes, the swallowtails were crowded together with wings upright and as tightly packed as plates in a rack. They were sipping water with their extended thread-thin probosces while at the same time excreting drops of it from the tips of their abdomens. It wasn't the water that they were seeking. They were absorbing the salts that were dissolved in it. And when I slowly lowered myself and sat down beside them, they settled on me too – on my forehead, my cheeks, my arms, delicately tickling me as their probosces probed over my skin. It was rather nice and I tried to sit still so as not to disturb them. But that was not easy, for while they were absorbing my salty sweat, large white-headed mosquitoes settled among them, and started drinking my blood.

That night they disappeared into the forest, presumably to roost. But the next morning they were back in the clearing as thick as ever. The blizzard continued for another day and then suddenly they were gone except for a few rather lost-looking stragglers, and some broken bodies on the mud down by the river, where a wave had washed over them. Perhaps there had been a sudden change in the weather – an increase in the temperature or humidity that I hadn't noticed – and that had brought about a mass hatching of all these different species in the area. Maybe one particular kind of tree, common in the surrounding forest, had just come into bloom and its perfume had attracted these millions. Maybe they were all on their way somewhere else in a mass migration.

For some butterflies, frail though they may seem, are – amazingly – valiant travellers. The most famous of these migrants are the monarchs of North America. As caterpillars they live in the country around the Great Lakes, in Minnesota, Michigan and Ontario, feeding on milkweed, a plant that contains a powerful poison that protects it from most other plant-eaters. Several generations of monarchs may hatch here in a single season. But as the days begin to shorten with the arrival of autumn, the monarchs begin to fly south. They travel by day and roost each night. As they go their numbers increase until eventually they form great processions. People in cities lying on their regular traditional routes welcome them with carnivals. Some among these travellers will stop in the southern states and find roosts in which to spend the winter. But most continue south.

These processions travel right down across the United States to Mexico and eventually converge on just one or two secluded valleys in the mountains. There they settle on a few groves of pine trees, covering the branches so densely that the pine needles beneath them are quite invisible. There are tens of millions of them. They do not feed. They are sustained entirely by the fat derived from the milkweed plants that they consumed as caterpillars far away to the north. Occasionally a few will flutter down to drink at a stream, but most simply hang there on the branches, as thick as fur. It is one of the most impressive sights in the whole of nature.

In spring they set off northwards. The majority stop to breed in the southern United States and then die, leaving their offspring to continue the journey to the north where their parents had hatched as caterpillars. So any one individual does not necessarily make the whole journey, there and back. It is the monarch as a species that performs this huge migration.

But why? We can only speculate, but it seems that the original home of both the monarch and the milkweed was in South and Central America. About two million years ago, the climate of this part of the world began to warm. The milkweed began to spread north with the warmer weather and the monarchs followed. But the change in the climate also brought drier conditions to Mexico and as a consequence, the monarch's food plants that grew there began to disappear. So the butterflies had to fly farther north to find places where it still flourished. At the same time, the winters in these new feeding grounds in the north became unendurably cold. So each autumn the butterflies flew back south to their ancestral home in the warm valleys of Mexico to hibernate until conditions in the feeding grounds to the north improved once again.

Sadly there are far fewer butterflies in my London garden these days than there used to be. Once my buddleia bush was covered each summer with red admirals, peacocks, painted ladies and tortoiseshells. Now the appearance of just one of these is enough to call me outside. I can only rely on the appearance of a cabbage white. Not so long ago I was guilty of taking little notice of them. Maybe I even dismissed them as pests, for indeed they can be. Now I am glad to see them.

If you do have cabbages, the cabbage whites are likely to spend some time with you fluttering around your garden. They may be courting. Some may even settle on a leaf and copulate. The females will then lay their egg in clumps on your cabbage leaves before fluttering away.

There's not really enough space for cabbages in my garden. At any rate, I don't grow any and the cabbage whites that visit me behave rather differently. They appear over my garden wall always from roughly the same direction. They are also likely to be flying in a much more purposeful way and after fluttering about for a few minutes, they disappear at some speed, continuing their journey in approximately the same direction. That will be the pattern for the first part of the summer as generation succeeds generation. But suddenly – at the height of midsummer – there is a change. That date is remarkably predictable. In the north of England it takes place in the second week in August. In southern England it happens a little later during the last week of that month. And in central France, it is later still. The effect of course is that, one way or another, cabbage white caterpillars will be nibbling away on the leaves of cabbage plants, not to mention broccoli, cauliflowers and Brussels sprouts, from the far north of England as far south as Spain.

How do they find their way? They only make these migratory journeys during the middle of the day when the conditions are at their warmest. So when any one butterfly sets off, the summer sun will be in roughly the same position and the butterfly is programmed to fly at the same angle to it that its parents did.

There is not much doubt now that our climate is changing. It is getting warmer. There are predictions that over the next few decades, southern Europe will become drier and drier. Maybe eventually northern France and England will no longer have a climate suitable for growing cabbages and brussels sprouts, while the warmer summers in Scotland will enable people there to grow them for longer. But even though the winters may be warmer, they will still be dark and plants will not be able to flourish as long as they do in the south. So the cabbage whites won't be able to stay in northern Scotland all year. They will have to leave and fly southwards perhaps to roosts in the pines and fir trees of southern France. There may be groves down there where the trees will be draped with millions of roosting butterflies, just as there are today in the mountains of Mexico. Will we then greet them on their way across England with Cabbage White Carnivals from Manchester to Milton Keynes? I would rather like to think we might.

*The pursuit of butterflies became a widespread enthusiasm among all sections of society during the nineteenth century. Perhaps the most famous and almost certainly the most extreme example was Lord Walter Rothschild who collected, mounted and housed over two million specimens. In the twentieth century, however, the increase in the size of the human population and the consequential reduction in wild places where butterflies might flourish has become so great that in Britain, sadly but necessarily, catching them is now illegal.*

Above. *Heliconid butterflies settle on my hand and drink my perspiration as I sit beside a river in the Paraguayan forest. Behind, closer to the river, clouds of yellow and white butterflies suck up mineral-rich moisture from the mud. A large mosquito also has taken the opportunity to collect a little blood.*

Right. *Monarch caterpillars feed for weeks on the leaves of milkweed, which are poisonous to many other animals.*

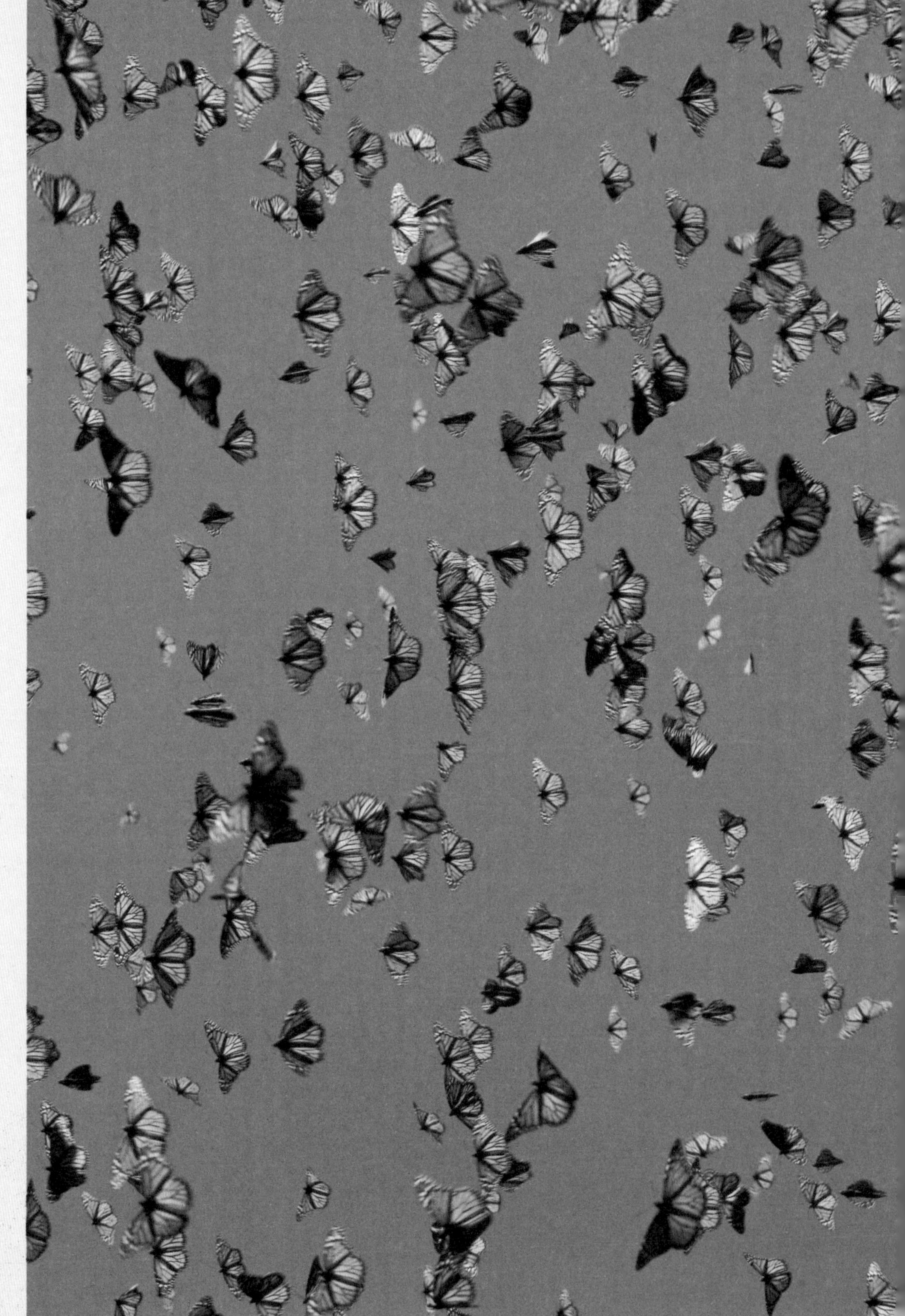

*Hundreds of thousands of monarchs take advantage of good weather and continue their migratory journeys together.*

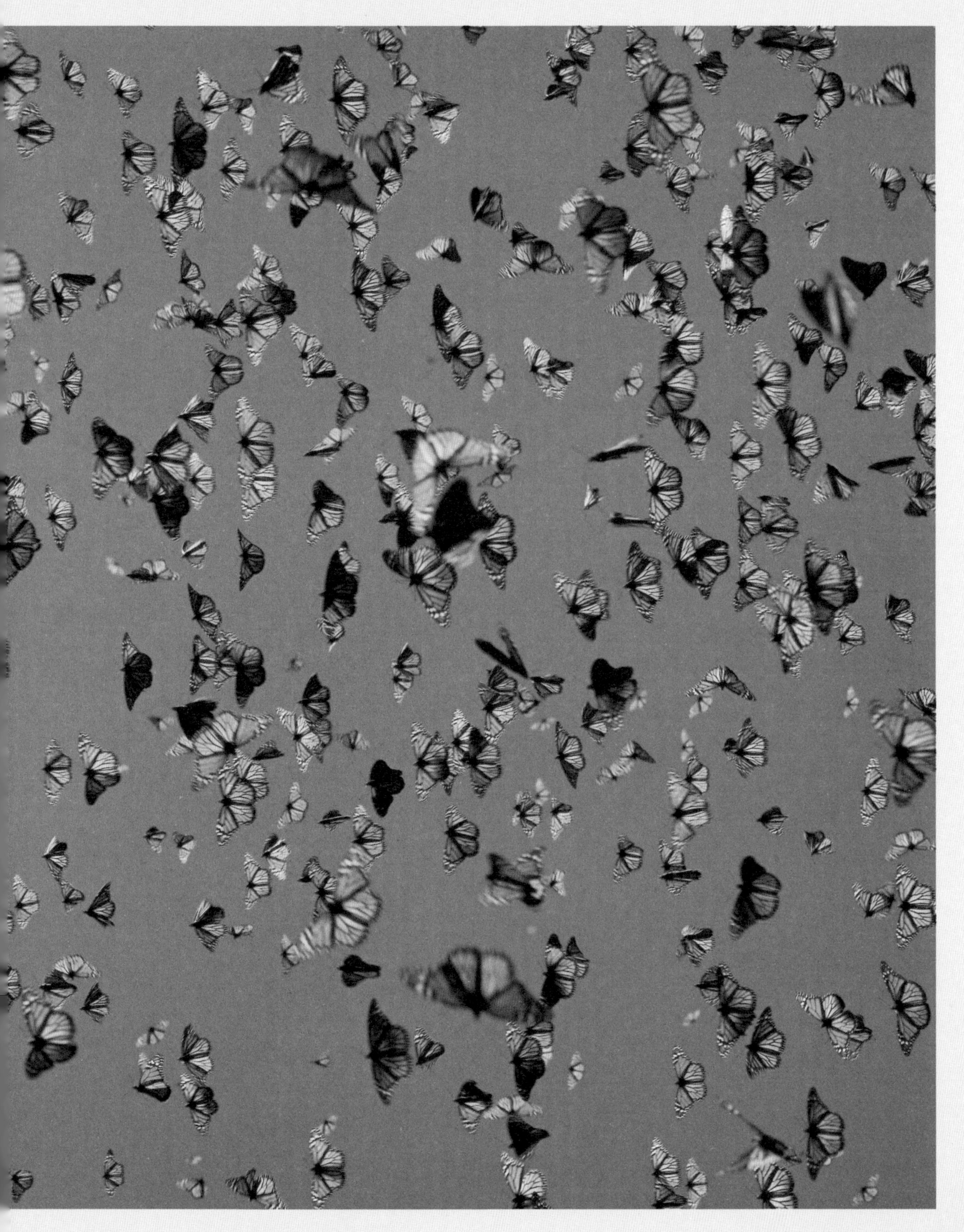

*The female cabbage white butterfly lays her eggs on cabbages and other plants of the* Brassica *family. She produces them at a rate of about four a minute and can lay as many as a hundred. With each batch she adds a chemical substance that is easily detected by other females and deters them from laying on the same plant.*

Right. *The caterpillars of the cabbage white work as a team, all feeding or resting at the same time. As they feed, mustard oil from the Brassica leaves accumulates in their bodies and becomes so concentrated that neither birds nor mammals such as mice or shrews will eat them.*

Below. *The final result of a visit by a female cabbage white.*

# 13

# Chimps

They say that chimpanzees and human beings share 99% of their genes. The precise figure varies but the implication is usually the same – that chimps are 99% human or we are 99% chimpanzee.

◁ *Chimpanzees are almost as varied facially as human beings. It is certainly possible to recognise individuals just as easily. They do not have as many muscles in their faces as a human being, but even so they clearly reveal their emotions with their expressions.*

They say that chimpanzees and human beings share 99% of their genes. The precise figure varies but the implication is usually the same – that chimps are 99% human or we are 99% chimpanzee. But that is not really an accurate way of interpreting those percentages. It suggests that the genome – the genetic make-up – of a species is rather like the recipe for a pudding. A little more of this or that will change the taste or consistency a little this way or that way. The fact is, of course, that a genome is much more like a programme tapped on to a computer keyboard. The number of keys you touch is not necessarily important. It's what they are that matters. Tapping one may make very little difference to the image on the screen, whereas a tap on another can totally transform it – and it is the presence or absence of those crucial genes that can produce the major differences – or similarities.

Nonetheless, our genetic makeup certainly *does* indicate that we are closely related to chimps. So what are these close relatives of ours really like? Well, our understanding of them has changed hugely over the past fifty years. Back then most people's ideas about chimps came from the chimpanzee tea-parties that used to be staged by zoos – and chimps were perceived as engaging, mischievous and rather naughty little children, stealing one another's cakes and throwing the crockery about. That was because the chimps that took part *were* children – chimpanzee children – only a few years old. But as they grew up and became bigger and stronger, they became less and less controllable, so they no longer took part in the tea parties. And we were not entertained by watching the surly, indeed rather alarming adults that, we now realise, were in truth prisoners with all the neuroses that such a deprived existence brings with it. Today, those tea-parties are a thing of the past, and adult chimps are given properly spacious enclosures with the logs and ropes and other things that they enjoy using so they can live, if not natural lives, at least fuller ones.

The earlier misconception about the true character of chimps was transformed by one truly remarkable woman – Jane Goodall. Fifty years ago, against all advice, she decided to study wild chimps by what was then the novel and courageous technique of simply going into the wild and sitting down and watching them. Many other scientists have now followed where she led. As a consequence our knowledge of the true nature of chimps is now hugely detailed and we recognise that they are intelligent, complicated, ingenious, powerful animals with marked individual personalities.

Among those scientists who followed her lead is a Swiss zoologist, Christophe Boesch. The chimps he works with do not live in open country as Jane's do, but in the dense forests of the Congo in west Africa. He was one of the first to show that chimps are not the amiable vegetarians that many had thought. They also eat meat. And what is more, on occasion, in order to get it, they hunt other primates – monkeys.

Twenty years or so ago, I spent some time, together with a film crew, trying to get pictures of this hunting behaviour which at that time had never been recorded on film. Christophe had started his study ten years earlier, simply by standing silently near a chimpanzee group and following them wherever they went in the forest. He did so not only day after day but year after year. Eventually one group became so accustomed to him that they took no notice of him whatsoever. He agreed that we might accompany him, provided that we wore exactly the same sort of clothes as he did and behaved in the same sort of way. And astonishingly, the chimps accepted us too. That is to say, they ignored us.

But it was an exhausting business. Christophe stayed beside his chimps every day wherever they went until, in the evening, they began to prepare for the night by climbing up into the tree tops and breaking over branches to make platforms on which to sleep. He couldn't leave them until they did that, because otherwise he would not know where to find them the following morning. If he lost touch with them, it might take him several days to find them again. Continuity would then be broken, essential data lost. So that meant we could not start to walk back to camp until the evening and sometimes did not get back until well after dark. And then the following morning, well before dawn, we would have to set off again in order to get back to the chimps when they were just getting up and before they set off for the day's activities.

So we had to do the same. Day after day. I must admit that I have never been more exhausted, before or since. Our main objective was to film them hunting, but that was not a daily activity. They only did so about once every ten days. Their targets were colobus monkeys. Catching them was not easy. A colobus weighs much less than a chimp so it can usually escape along branches that would break under a chimp's weight. By itself a chimp can seldom corner a monkey. That needs teamwork.

It is easy to see when a hunt starts – even if it is not so easy to work out how the group arrives at the decision to do so. Several of the senior males start to stride through the forest in a strikingly meaningful way, visibly different from the rather more haphazard roistering that normally goes on. The females and youngsters follow behind. They too are visibly excited. They can hear the calls of colobus in the distance. Before long the hunters split up. One or two of them move quietly ahead making a detour so that they get beyond the tree where the colobus troop is feeding. Then quietly and unobtrusively they clamber up into the canopy and settle down there, partly shrouded by leaves. They have laid an ambush.

Now the other hunters, taking little care to conceal themselves, run up the trees or climb up lianas hand over hand towards the canopy. The colobus see them and take fright. There is one obvious escape route through the branches ahead – and it leads straight into the ambush. Out come the hidden chimps. The colobus group scatters, shrieking with terror and most escape. But one female, encumbered by a baby clinging to her back, takes a wrong turn and in attempting to escape her pursuer, she jumps almost directly into the arms of the ambusher. She and her baby are grabbed.

Down below on the ground there is pandemonium. The female chimps and youngsters are shrieking with excitement as the hunters descend, carrying the dangling bodies of their victims. By the time we catch up with them, they are sitting on the ground in a group and fragments of the monkeys' bodies – arms, legs, heads – are being passed around and stripped of the meat.

I sat down, exhausted by the chase, chest heaving and drenched with sweat. And it was, of course, a horrific sight as the all-too-human looking fragments of the little bodies were torn apart. But as I watched, two things were clear. First of all, the hunt I had watched had depended on team-work. Those taking part in it had understood one another. Each had known what his job was. Secondly, and perhaps even more interestingly, there was a considerable degree of unselfishness among the group. The individual who had made the actual kill did not keep it all to himself but allowed others to take a share. In fact Christophe told us that sometimes, if the victim was small, the hunter himself did not get any of it at all.

Maybe our distant ancestors also behaved in a similar way. We can be pretty sure from the shape of their teeth and guts that they like chimps had a mixed diet. And they were clearly not powerfully armed with huge dagger-like teeth and great claws. Nor were they swift enough to run down their prey. They too, if they were to be successful hunters would have had to be cooperative and unselfish.

It may or may not be that the source of this drive to hunt, this savagery, is in one of the 99 odd per cent of genes that we share with chimpanzees. The thought may shock us, but we ought at least to recognise that it was this need and passion for the hunt that led to the development of communication, team-work and cooperation and that, paradoxically, it was this that ultimately led the human species to some of its greatest achievements.

*Early naturalists were not at all certain about defining humanity. Edward Topsell, the sixteenth century naturalist included this picture in his great natural history and firmly labelled it an ape. It was probably based on a chimpanzee. By the end of the twentieth century, however, some scientists were suggesting that, to the contrary, human beings should be regarded as a species of chimpanzee.*

Above. *Jane Goodall revolutionised our understanding of apes when she boldly lived alongside them in the wild for so long and with such patience that they eventually totally accepted her.*

Below. *Jane's studies revealed the complex detail of chimpanzee family life and the degree to which young chimpanzees are cared for and learn from their parents.*

Above. *Chimps crack nuts using a small stone as a hammer and a larger flat one as an anvil. Good anvils are quite rare and after collecting nuts chimps may wait in a queue to use a favoured one.*

Right. *Termites will bite a stem of grass poked into one of their holes, and so can be fished out and eaten. Baby chimps sitting beside their mothers watch intently and learn these techniques.*

Above. *Chimp hunters, having caught and killed a red colobus monkey, eat some of the meat immediately but will take most of the body down to the females and young who are waiting on the ground for a share.*

Below. *A high-ranking male chimp claims the much-relished carcass of a red colobus monkey.*

# 14

# Cuckoo

Everyone, I suppose, knows what is meant when you talk about a cuckoo in the nest. It means someone, or something, that has appeared in a place where it shouldn't, and has pushed out the rightful owners.

◁ *A parent reed warbler, undeterred by the replacement of its own nestlings by a single alien many times their size, continues to feed the imposter.*

Everyone, I suppose, knows what is meant when you talk about a cuckoo in the nest. It means someone, or something, that has appeared in a place where it shouldn't, and has pushed out the rightful owners. And that is accurate enough. A female cuckoo does lay her egg in a nest built by other kinds of birds, and her young chick does oust the egg or the chick of the nest-builders which then look after the intruder.

But no bird in Britain has caused more puzzlement and fury among ornithologists than the cuckoo. One argument which raged back in the 1920's was about the way that a female cuckoo manages to get her egg into another bird's nest in the first place. It doesn't seem to be much of a problem in most instances, but although it happens rarely some cuckoo eggs appear in small domed nests like those of a wren, or ones with tiny openings, like those of a swallow. A female cuckoo who is in all cases significantly larger than the bird she parasitizes, is too big to squeeze into such nests. Some experts back then maintained that she first lays her egg on the ground and then picks it up in her beak. Holding it either in her beak or her crop, she flies off to another bird's nest and drops or regurgitates her egg into it. As evidence they pointed out that female cuckoos are regularly seen flying with an egg in their beaks.

Other experts argued that in such instances the female is carrying not her own egg but one of her host's which she is taking away so that the number of eggs in the nest in which she has just deposited hers, remains the same and the rightful owners won't notice. As for laying, they maintained the bird did so directly either squatting on the nest or clinging to the rim, facing outwards. This view was held by, among others, an expert ornithologist called Edgar Chance. He was a fanatical egg-collector – the practice then was legal – and sufficiently wealthy to be able to pursue his hobby on a grand scale and in 1922, he commissioned one of the first really great natural history film-makers, Oliver Pike, to make a film that would settle the question one way or the other.

The combination of Pike's ingenuity in building hides to accommodate his huge 35 millimetre movie camera and Chance's knowledge of exactly where to place it, produced a truly epoch-making film. It was made on some heath-land in Worcestershire and it showed a female cuckoo flying into a meadow pipit's nest, squatting on it – in spite of being fearlessly harried by the pipit parents – spreading her tail and after a matter of a few seconds flying off, leaving behind her egg in the pipits' nest. That, as far as I know, was the first natural history film ever to resolve a scientific argument.

Even so it wasn't the complete answer. How *did* a cuckoo get her egg into a nest with a tiny entrance? It eventually turned out that a cuckoo's cloaca – that is to say the aperture beneath her tail from which her eggs come – is muscular and extendable. So a female cuckoo doesn't so much deposit an egg as *squirt* it into a nest – and that can be done while clinging to the rim.

But there was another mystery about cuckoos. They lay their eggs in the nests of several dozen different species, but eighty percent of them parasitise just three species – reed warblers, meadow pipits and dunnocks. The colour of cuckoo eggs varies, for each approximates to that of the host species. Those inserted into a pipit's nest will be mottled brown like a pipit's, those in a warbler's nest greenish white, spotted and blotched with a darker gray. That has to be so because if an egg is markedly different in colouring or size, a pipit or a warbler will quickly either peck a hole in it or throw it out.

Does that mean that an individual cuckoo can vary the colour of her eggs? No. It means that there are different groups within the single species of British cuckoo, each of which habitually parasitizes only one species of host. The relationship between the cuckoos and these species has existed for so long, with the hosts throwing out obviously alien eggs, that now each clan of cuckoos lays eggs that match those of their particular victims.

There is an exception to this. A dunnock's eggs are pale blue but the cuckoo's egg that sometimes appears among them has dark spots on it. Yet in most cases, the female dunnock doesn't seem to be upset by this. Is that because she hasn't noticed? It may well be. It seems likely that cuckoos only picked on dunnocks comparatively recently – and dunnocks haven't yet woken up to what is going on. Maybe in time they will become critical enough to identify a strange egg and start throwing it out, just as pipits and warblers do. Then the cuckoos will either have to evolve a matching colour to their eggs – or pick on another species.

The cuckoo family is a large one. There are some 139 species world-wide. Many are perfectly responsible parents incubating their eggs and caring for their young in the normal way. But some aren't. In South Africa, there is one, the diederick cuckoo, that parasitises weaver-birds. They do something that makes life quite hard for the diederick. Their eggs vary considerably in colour, from one individual to another. And what's more they weave nests that are globes the size of a grape fruit with an opening on the side so shaped that you can't see within, even if you are perched on the entrance. So a female cuckoo has no way of selecting a nest with eggs that match the colour of the one she is about to lay. She may fly in, deposit her egg with the speed and aplomb typical of a cuckoo, only to find that the weaver-bird immediately detects what has happened because the new egg is the wrong colour and throws it out.

If however a cuckoo's egg remains undetected, things soon turn nasty. The female picks the moment to lay her egg very precisely. She does so early in the period in which her host is herself still laying, and her chick hatches very quickly so her nestling appears first. And when it does hatch, it really is a monster, naked, with lids still closed over its huge bulging eyes, a pair of unfeathered wings, like arms, and a hollow in its back between what you might call its shoulder-blades. It manoeuvres itself underneath each of its host's eggs or newly-hatched nestlings and, one by one, heaves them on to its back and pitches them out.

It then demands food with begging cries which are so urgent, loud and frequent that its foster parents labour incessantly, stoking it with insects of one kind or another. It grows with astonishing speed. Within days it is bigger than its foster parents and yet still it demands – and gets – food. By the time it starts to fledge it will have consumed as much food as ten reed warbler chicks.

And now comes yet another cuckoo mystery. How is it that parent birds that are so extremely particular about the size and colour of their eggs take no notice of the fact that the insatiable creature they are feeding is so grotesquely different from one of their own nestlings in both general appearance and size? It has been discovered that the calls it makes are similar to those of the nestlings it has displaced, except that they are more frequent and more insistent. But the young bird that now occupies the nest is nonetheless several times their size and looks nothing like them.

And there is one final mystery. By the time the young cuckoo is fully fledged and ready to fly, its parents have long since left Britain and are already back in Africa. They stayed here for only a few weeks – the shortest visit to these shores by any migrating bird – leaving their offspring still being cared for by pipits, warblers and dunnocks. So the young cuckoo has no adults of its own kind to guide it or accompany it. How is it that nonetheless it manages to find its way across the Sahara to the very forest or savannah where its parents came from?

So there are still *some* cuckoo mysteries. It would be nice to think that natural history film-makers today could help in solving them just as the great Oliver Pike dealt with that one back in the 1920's.

THE CUCKOO.

*A calling cuckoo as portrayed by the great Newcastle wood engraver, Thomas Bewick, in his 1804 History of Birds. Unlike most bird artists of his time and before, he chose to work, whenever he could, not from skins of dead specimens but by watching the living bird.*

Above. *A cuckoo making its celebrated two-note call, to attract a mate.*

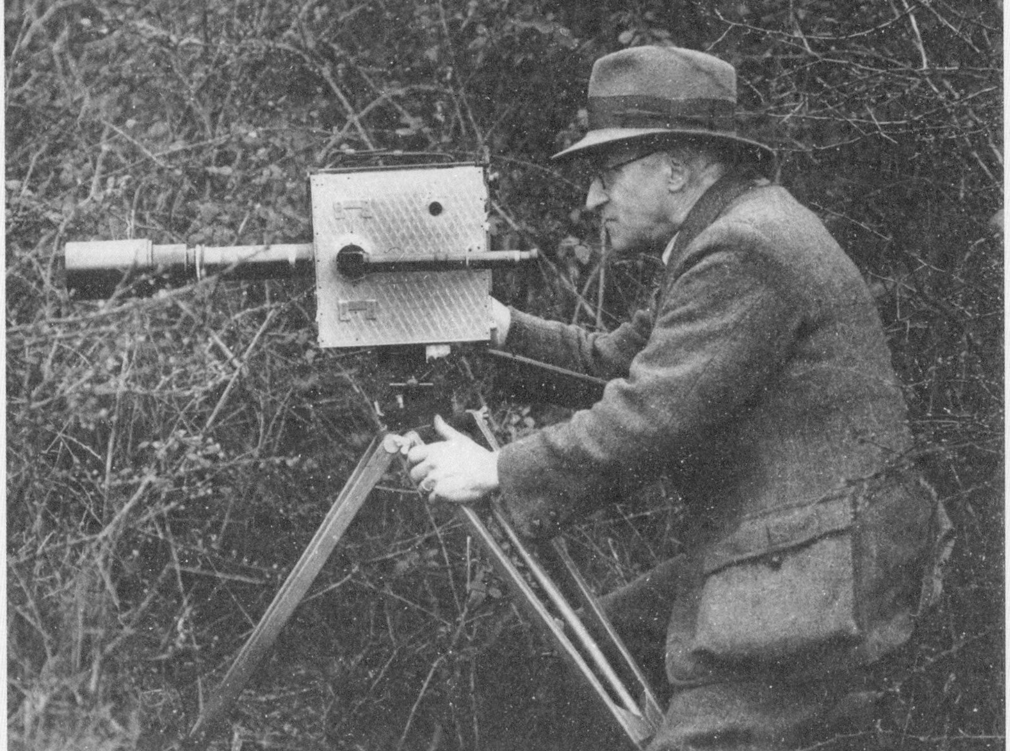

Right. *Oliver Pike, the pioneer natural history film maker, with his hand-wound 35 millimetre ciné camera with which he filmed the first-ever photographic record of a female cuckoo laying an egg,*

Above. *The largest egg in this thick-billed warbler's nest, in eastern Russia, was laid by a cuckoo. Its close resemblance to the warbler's own eggs suggest that the parasitic relationship between the two species is a long-established one.*

Below. *Dunnock's eggs are pure blue. The cuckoo's egg sitting alongside them is very different. Yet dunnocks, in whose nest such things appear, seldom seem to notice or to care.*

*All the eggs in this reed warbler's nest are similarly coloured and patterned. But one is slightly paler and larger. It was laid by a cuckoo.*

*The biggest egg developed particularly quickly and has hatched first. Now the young cuckoo, its eyes not yet open, begs for food.*

*The blind featherless chick almost immediately proceeds to push out its hosts' eggs..*

# 15

# Quetzalcoatlus

The biggest animal to fly was not a bird. It was a reptile. Just how big it was we can't be sure – yet. But I'll come to that later. The first example of its kind ever recognised was not, in fact, very big – only about the size of a thrush.

◁ Quetzalcoatlus *species were among the last members of the pterosaur dynasty. We cannot be certain of their colour, for pigment does not fossilise, but there is evidence that their bodies were covered with short fur, and their bones prove that their wings stretched nearly forty feet, tip to tip.*

The biggest animal to fly was not a bird. It was a reptile. Just how big it was we can't be sure – yet. But I'll come to that later. The first example of its kind ever recognised was not, in fact, very big – only about the size of a thrush. The great French naturalist, Georges Cuvier, found its fossilised bones in 1801 in one of those cabinets of curiosities that European aristocrats used to assemble. It had been thought to be a fossilised bird, but it was very clearly not a true bird, because instead of a beak it had bony jaws lined with teeth. Cuvier decided that it was a reptile. Its arms, however, were very odd. Each hand had four separate digits but one of these, the fourth, was hugely elongated. Cuvier realised that this must have supported a skinny membrane attached somehow to the animal's body and that therefore the whole arm functioned as a wing. So he called this strange creature a pterodactyl – using two Greek words that together mean 'wing finger'.

A few years after that, in 1828, Mary Anning, a famous fossil collector living in Lyme Regis in Dorset, found something similar. Her specimen, however was even more reptilian than Cuvier's for whereas his pterodactyl was practically tail-less, Mary's animal had quite a long bony tail which, as we now know, supported an extension of the main wing membrane that stretched between the animal's two hind legs.

But Mary was also discovering all kinds of other wonders in the Dorset cliffs. There were huge swimming reptiles – ichthyosaurs and plesiosaurs – including some that were forty feet long. And then around the same time, a doctor from Brighton, Gideon Mantel, discovered some strange fossilised teeth in a quarry on the South

Downs. They looked exactly like lizard teeth except that they were, in comparison, gigantic. Dr Owen, the Director of the Natural History Museum, decided that they had belonged to an immense reptile as big if not bigger than an elephant. He called these new creatures dinosaurs – that is to say 'terrible lizards' – and people came to realise that once, millions of years ago, the Earth was ruled by giant reptiles.

These land-living monsters dominated people's ideas about that far distant time, but new kinds of winged ones were also being discovered and they were sufficiently different from Cuvier's for a new name to be needed that would include them all. So the whole group were called pterosaurs – 'flying reptiles'.

And very varied they were. Some had pointed dagger-like teeth suitable for snatching fish from the surface of the sea. Some, instead of teeth, had rows of dense bristles that may have been used to filter out small crustaceans or such-like things from water, rather as flamingos do today; and some had long bony jaws with no teeth at all that may have been used as forceps to pick up edible bits and pieces, perhaps even carrion. But none of these were giants comparable in size to the giant dinosaurs.

And then, in 1971, a young research student, Doug Lawson, working in the desert lands in the south of Texas found a fossilised bone. He knew that it was a pterosaur bone. Flying animals – like airline passengers today – have to keep their weight to a minimum and for the same reason. Pterosaur bones therefore are not only hollow, but have walls that are paper-thin, supported by an internal network of crisscrossing struts. Birds today have similar hollow bones, but the walls are not nearly as thin as those of a pterosaur. So Doug was quite sure that he had found the bone of a pterosaur. He also knew from its shape what particular part of the body it came from – the wrist. But whereas in most pterosaurs this particular bone is the size of a grain of rice, this one was as big as a grapefruit. The animal that it had belonged to must have been a monster. He and his professor, Wann Langston, started an intensive search. They did find other bits and pieces, but no skull, no backbone, no other remnants that could provide irrefutable proof of the exact dimensions of their newly discovered aerial giant.

Nonetheless there was enough for the animal's existence to be published – and for Doug to exercise his right, as the first discoverer and describer, to give it a name. And he thought up a real tongue-twister. He knew that the Aztec people who once lived a little farther south in Mexico, worshipped a god that they believed to be a flying reptile – a feathered serpent to be precise – and the parallel appealed to him. The Aztecs called this god Quetzalcoatl, so Doug called his discovery *Quetzalcoatlus*. And he calculated that, if the animal's body had the same proportions as other pterosaurs, then judging from the wrist bone it must have had a wingspan of nearly forty feet.

This caused something of a stir. Some palaeontologists said that such scaling up was unjustified. Nonetheless, Doug stuck to his guns. And in due course whole wings were found to prove that this amazing creature did indeed have a wingspan of such immense dimensions. In the last twenty years, lots more species of similar giants have been found not only in Texas but in Russia and Brazil.

They are astonishing creatures – and not just because of their great size. They had the most extraordinary necks. There are five bones in them, as there are in other pterosaurs, but some of these are enormously elongated. Not only that but they fit together so neatly and tightly that the neck itself is almost as stiff as a pole. What would a huge flying reptile do with a neck like that? Well, *Quetzalcoatlus* didn't have any teeth. It seems it was a scavenger, feeding on carrion. I have watched vultures at work on the carcass of a big antelope. They stick their heads deep into the body of the corpse they are feeding on. That's why they have naked necks. Feathers would get horribly filthy and matted with blood. And what kind of body did *Quetzalcoatlus* feed on? Why, the mountainous corpses like those of *Tyrannosaurus* and *Gigantosaurus*. To get into the body cavities of one of those monsters, you would need a special tool. A stiff rod-like neck the size of a giraffe's would do very well.

But these great flying giants were the last of their kind. The pterosaur family as a whole had been declining in both numbers and variety for the previous fifty million years or more. That might have been because of competition, for another branch of the reptiles had also taken to the air. These were the early birds.

Birds have feathers and those in their wings were long and strong. The membranes of the pterosaur wings, on the other hand, had to be anchored at the back if they were to catch the air as indeed they were, first at the ankle and then, in later forms, around the knee. So whereas birds, searching for food, could run and hop, patter about over sand or wade in the shallows, pterosaurs were to some degree hobbled and at a disadvantage. But whatever the reason, the fact is that birds increased in number and variety while pterosaurs decreased. And when, for whatever reason, the dinosaurs disappeared, the pterosaurs went with them.

But what a sight they must have been – with wings four times the size of any bird flying today. Some time ago, I was making a programme about them during which we filmed a full-scale radio-controlled model gliding backwards and forwards above the very cliffs in Dorset where Mary Anning had first found their remains. It wasn't quite the size of *Quetzalcoatlus* but based on a rather smaller species with wings a mere 15 feet across. But it looked magnificent. After finishing our filming, I had to get on a train and dash back to London to attend a reception that was being given by a conservation charity. It was a rather posh occasion and I had to change into a dinner jacket and black tie on the train. As I went into the crowded room I was met by my hostess. 'How lovely to see you,' she said, 'And what wonderful creature have you been filming for us today?' Here was my chance to really impress. 'Well.' I said, 'As it happens I have been filming a pterosaur gliding above the cliffs of Dorset'. 'Oh,' she said' they *are* so lovely, aren't they' – and turned away.

For a moment, I felt a little put down. But come to think of it, she was absolutely right. They are – or rather – sadly, they were.

*Quetzalcoatl, the plumed serpent god of the Aztecs after whom the biggest of all pterosaurs was named, here shown by a 16th century Aztec scribe consuming a human being.*

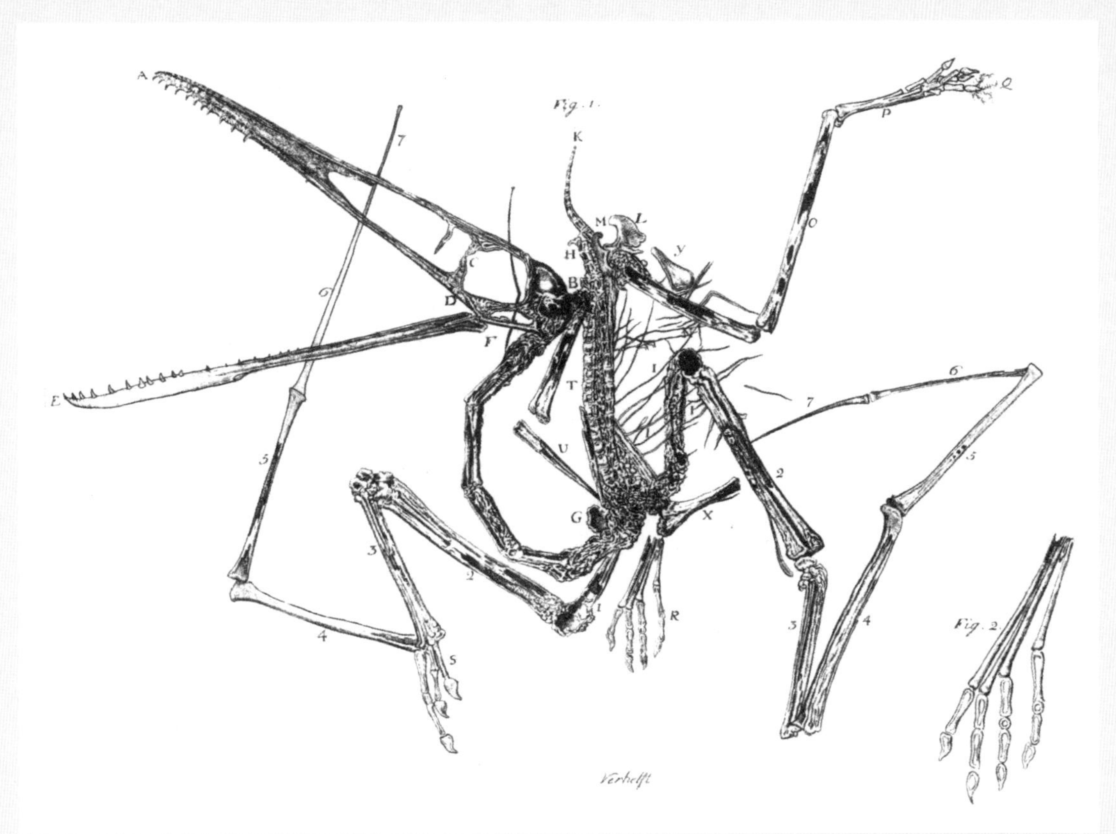


Above. *A strange fossil was found at Eichstatt in the eighteenth century and kept in the cabinet of curiosities assembled by the Elector Palatine in his palace at Mannheim.*
*This is the meticulously accurate engraving that accompanied the 1784 announcement of its discovery.*

Right. *Georges Cuvier, the great French scientist and comparative anatomist sitting beside his microscope and bottled specimens. It was he who in 1801 was the first to recognise that the bizarrely long-fingered fossil at Eichstatt must have belonged not to some kind of sea creature, as had been suggested, but to a flying reptile.*

Above and right. *In 1985, Steven Winkworth, an expert builder and flier of model radio-controlled aeroplanes, built a full-scale model of* Pteranodon, *a pterosaur that had a wingspan of fifteen feet. The model was unable to flap its wings as a living pterosaur would have done and its feet, with toes wide spread, had to be used as rudders. Nonetheless, when Steven launched it from the cliffs in Dorset it glided back and forth most convincingly.*

Above.
*Quetzalcoatlus was almost certainly a carrion-feeder. Its huge neck, as long as a giraffe's, would have enabled it to probe deep into the carcasses of giant dinosaurs to pick out the entrails.*

Right. *Doug Lawson, in a Texan cabin, explains how, as a newly graduated student, he found a pterosaur bone so big that it that led him to believe it came from an immense pterosaur with a forty-foot wingspan.*

# 16

# Chameleon

We called him Rommel after the German general because he came from North Africa, where Rommel fought his most famous campaigns during the Second World War. I had brought him back from a trip to Algeria, carrying him in a little cardboard box

◁ *Chameleons do not grow more than a foot or so in length but when angered or threatened they hiss, inflate their bodies, and darken in colour in an alarming way. It is not surprising that many people believe them to be dangerous. This species with three horns on its nose is from West Africa.*

We called him Rommel after the German general because he came from North Africa, where Rommel fought his most famous campaigns during the Second World War. I had brought him back from a trip to Algeria, carrying him in a little cardboard box – which was a perfectly legal thing to do at the time. He was a chameleon. He wasn't the first of his kind to make the trip. There are accounts of soldiers in the Roman Army returning from North African campaigns bringing similar strange little lizards back to sunny Italy to astonish the folks back home.

And chameleons really are astonishing, with their feet like forceps and their independently swivelling eyes. Not least unsettling about them is the fact that they are quite incapable of moving at any speed, but normally progress very slowly, carefully lifting one foot after the other as they clamber around in the branches of bushes.

The most wonderful thing about them is the way they eat. Rommel would obligingly demonstrate this whenever he had the chance. We used to put a stick in front of him on to which he would climb rather groggily, swaying somewhat from side to side, and then we would take him out into the garden and aim him, like an animated pistol, at a blue-bottle. He could recognise one from a good eighteen inches away. You knew immediately when he had noticed it. His eyes would swivel until both pointed at the fly straight ahead, so giving him a binocular view of it and thus the ability to accurately judge how far away it was. Sometimes when taking aim, he would also sway backwards and forwards or from side to side, the equivalent, I suppose, of fiddling with the focus on your camera to make sure you have got the picture sharp.

When he considered his prey was within range he would slowly open his mouth, exposing its bright yellow lining and toothless gums, and then bring forward a lump of muscle that lay in the floor of his mouth and contained within it a long tapered bony rod. As he opened his mouth a little wider and pushed the lump a little farther forward, he squeezed the muscle so that the front part of it slipped off the tapered bone and in a fraction of a second what was a short thick muscular tube became a long thin one, several times its original length.

Now it is customary to say, when describing feeding chameleons that their aim, in terms of both direction and elevation is 'unerring'. Rommel's wasn't, I have to say. Sometimes he fell short. Sometimes he missed to the side. But usually he got it right. The end of his tongue would hit the fly four-square and grab it. I say 'grab' because that indeed is what happened. The end of the tongue was swollen like a club but had a dent in the front with which he could literally grasp the fly. And then he contracted the muscle, pulling his tongue back into his mouth, bringing the fly with it. After that Rommel would chomp his capture with slow deliberation while his eyes started to swivel again in search of another target.

Chameleon eyes are unique and astonishing. They are conical in shape with only a tiny opening at the tip of the cone. And they swivel so that they can look not only vertically upwards and downwards, but backwards and forwards as well. And not only that. They operate independently so that while one is looking ahead, the other may be searching the branches behind. You might well wonder how on earth the chameleon's brain manages to deal with the two totally different and constantly changing pictures sent by its eyes. Apparently it accepts a signal from each one alternately, each lasting no more than a second.

Perhaps chameleons are most famous, not for their remarkable tongues or their amazing eyes but for their ability to change colour.

It's often assumed that they do this in order to match their surroundings and thus conceal themselves – hence the old joke about the chameleon that landed on a Scotsman's kilt and in consequence exploded. In fact while they do indeed match their surrounding to a certain extent, they do so more readily in order to express their emotions. If Rommel was irritated or felt threatened, he would go black with rage.

One morning when I took him out to give him his daily ration of blowflies, I was astonished to see on the twigs around him a dozen or so tiny replicas of himself. Rommel, apparently, was a female. I can be forgiven for getting his sex wrong for there is no obvious way to sex a chameleon.

But Romella, as we called her from now on, was unusual in giving birth to live young. Most species of chameleon lay eggs. Romella didn't because she belonged to a species that lives in the more northerly parts of the chameleon family's range where a female can no longer rely on the sun to provide sufficient warmth to hatch unattended eggs. So instead of depositing her eggs, a female keeps them inside her body and deliberately moves around to the sunnier spots to keep her body – and therefore her eggs – properly warm. She is a sort of mobile incubator.

However, we were now faced with the problem of what to feed to a dozen tiny chameleons. The answer turned out to be fruit flies presented individually to each one of them on the tip of a fine camel-hair paint brush.

The homeland of the chameleon family seems to be the huge island of Madagascar. At least sixty different species are found there – more than in the whole of the rest of the world put together. The smallest is really tiny – less than an inch long, smaller than many a grasshopper. It is probably just about the smallest of any land-living vertebrate but packed inside its minuscule body are all the organs a vertebrate needs – lungs, stomach, liver, kidneys. Its heart must be hardly larger than a pinhead. All in all it's a miracle of miniaturisation. And it has abandoned most of its family's normal characteristics. It doesn't live in bushes but down on the ground and its tail is not used for grasping but has become reduced to little more than a stump.

Nonetheless it is still capable of getting very angry. Pick one up and let it sit on the tip of your finger and it starts to vibrate. I suspect it is roaring as loud as its tiny lungs will let it, but the sound is so high-pitched that it is beyond the range of human ears. So to a human eye it just vibrates with anger.

The biggest of Madagascar's chameleons is a very different animal indeed. It is over two feet long, so big that it can – and does – eat not just insects but nestling birds. In colour it is very dramatic – green conical eyelids with blue stripes radiating from the tiny exposed eye, flanks with yellow stripes on a green background speckled with blue and crimson spots – an extraordinary palette which it can vary according to its mood.

Madagascan people, by and large, are terrified of any chameleon, maintaining that the mere flash of a chameleon's tongue is enough to blind you. Fifty years ago I was in Madagascar making films and collecting a few animals for the London Zoo. We were travelling from place to place, and of course we had to carry our animals with us.

One night someone broke into our car and in the process smashed the window of the driver's seat. That meant, of course, that we couldn't lock the car – and that was pretty serious since we had a lot of valuable filming equipment in the back. Among the live animals we had collected was a particularly splendid giant chameleon. So before leaving our car anywhere, we used to take it out of its cage and put it on the steering wheel which it grabbed firmly with its caliper-like feet.

And there it stayed. If anyone opened the door, this monster would turn and glare at them, swaying back and forth, opening its mouth, changing its colour before their very eyes and hissing loudly, gulping in air so that it literally swelled with rage.

Nobody interfered with our car after that. And to tell the truth, I didn't find it all that easy to get into it myself.

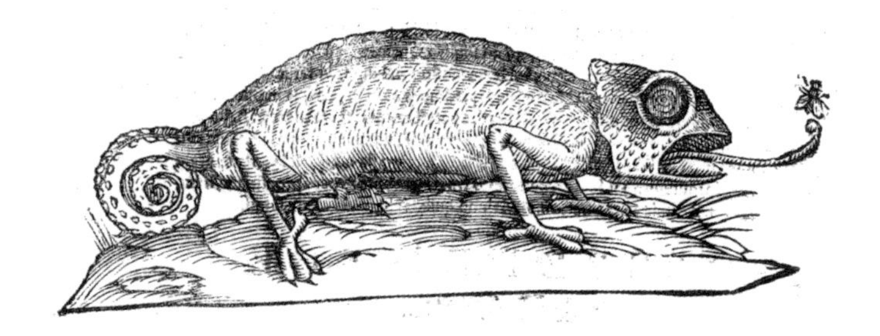

*The chameleons' heartland is Madagascar and Africa but one species has managed to cross the Mediterranean and can be found in southern Portugal, Spain and Sicily. It was doubtless the model for this illustration from the great encyclopaedia compiled by the Italian scholar Aldrovandus and published in 1599.*

*The chameleon's remarkable extendable tongue has a muscular end with which it grasps its prey.*

*The chameleon's tongue is tubular and has no rigidity except for a short section nearest its mouth where it is stiffened by an internal tapered bone. The rest of it is carried forward by the speed and force of its muscular contraction but then, having – usually – struck its prey, it is retracted into the mouth before it flops.*

Above. *Male chameleons defend their territory and quarrel over mates in a very aggressive way. But much is bluff and the loser retreats, sometimes by falling off the branch, before any damage is done.*

Right. *A chameleon's eye can swivel in all directions and moves quite independently of the eye on the other side of the animal's head. So a chameleon can see backwards and forwards simultaneously, should it wish to do so.*

Above. *A stump-tailed chameleon on a finger-tip. The members of this chameleon family are the smallest of all and totter about in the leaf litter on the forest floor. Since they have abandoned climbing, their tails have shortened and lost their grasping ability.*

Below. *Being so small and matching the leaf litter so closely in colour and outline, stump-tailed chameleons are easily overlooked.*

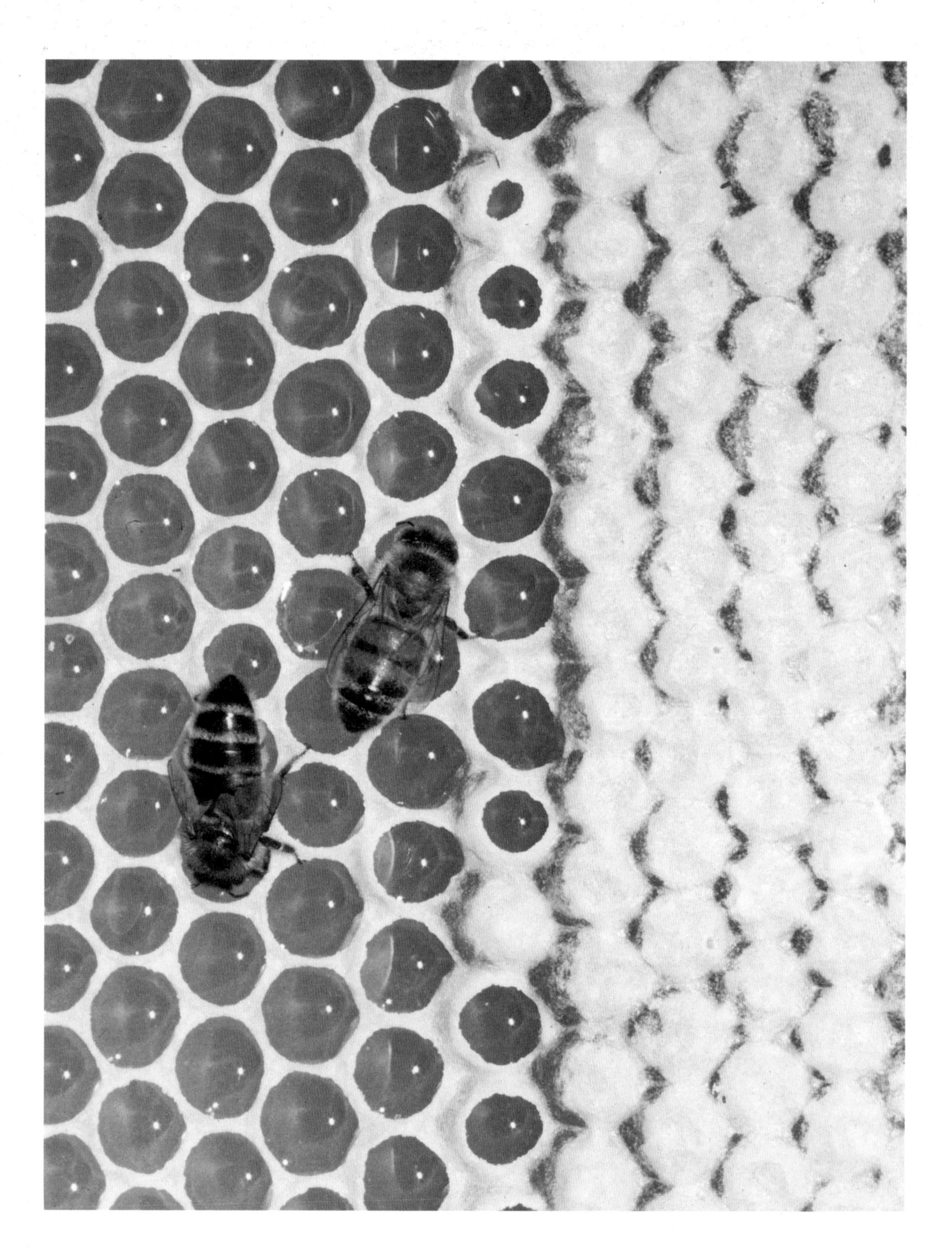

# 17

# Nectar

Nectar was almost certainly the first bribe in the whole history of life. It was without doubt on offer a very long time ago – around a hundred million years.

◁ *Worker bees turn flower nectar into honey by repeatedly drinking it from the cell where it has been deposited, adding digestive fluids from their stomachs and absorbing some of its water before regurgitating it into the cell. Here worker bees are tending a honeycomb. Those cells they have topped with a white cap of wax contain honey that has been fully processed.*

Nectar was almost certainly the first bribe in the whole history of life. It was without doubt on offer a very long time ago – around a hundred million years. We can deduce that because we have fossils of flowers dating from then – and flowers, of course, developed with only one function – to advertise the fact that they contained something worth having. Initially, that something may well have been pollen. It is very nutritious and doubtless many insects, then as now, eagerly ate it. But from a plant's point of view, pollen is quite expensive to produce. It contains a lot of protein. Nectar on the other hand, is relatively cheap. It is after all, little more than a solution of sugar in water. And it seems that insects liked it just as much if not more than the expensive pollen.

We have fossils of these first flowering plants. They were ancestors of today's magnolias. Judging from living magnolias, their nectar exuded from the base of their petals. The amount of nectar any flower produces is quite critical. There must be enough to make an insect's visit worthwhile; but not so much that it gets all that it wants. That way, when an insect has finished what's available in a particular flower, it will leave to look for more, carrying with it pollen grains that it has inadvertently collected as it brushed by the flower's stamens and that may be brushed off in the next flower it visits. If that happens, then the plant's bribe of nectar has worked. It has brought about cross-pollination.

The arrangement however is not as good as it might be because an insect visiting a magnolia often visits other kinds of flowers as well. The magnolia pollen it carries may be brushed off but it will not fertilise these other flowers – and that is a waste of valuable pollen. But before long – geologically speaking – some plants overcame this difficulty. They developed special nectar-producing glands – nectaries –

that were so placed and so shaped that only those species of insects that had evolved a particular kind of mouthparts were able to reach them. Insects that have such equipment will then tend to favour those flowers where they will be almost certain to find nectar since it is specially reserved and only available for them.

The most extreme example of this trick is a Madagascan orchid that has nectaries placed at the end of narrow tube-like spurs that are about a foot long and trail from the back of the flower. There is only one insect, a particular species of moth, that has a proboscis, curled up like a watch-spring beneath its head when not in use, that is long enough to sip it.

Most insects simply drink enough nectar to keep themselves going, but some live in places where there are seasons in the year when no nectar is available. Then they have to labour during the flowering season to build up reserves that will sustain them during the hard times. Ants in the Australian desert do that. They have a special caste of workers in the colony called 'repletes'. Workers, having gathered nectar from flowering gum trees, feed it to a replete which eventually takes in so much that its body swells to the size of a pea. Aboriginal people living in the desert are so knowledgeable and sharp-eyed that they can recognise the particular species of ant that does this. The honey-hunters then follow it to the little hole in the ground that is the entrance to the ants' nest. They dig and several feet down, in long horizontal galleries, they find dozens of repletes, hanging in rows. They are a lovely translucent amber colour with, at one end, a tiny head and minuscule legs with which they suspend themselves from the gallery roof. Workers will then come to them during the hard times, solicit and be given a little drink of nectar to keep them going. But take one of these repletes between your teeth, give it a little squeeze and warm marvellously sweet honey will spurt into *your* mouth.

The most dedicated collectors and storers of nectar, of course, are bees. A worker comes in with a crop-full and feeds it to another that is on house duties in the nest. This takes it and deposits it in a cell on the comb. It then fans it so that some of the water it contains evaporates. Then it sucks it back into its crop for a little more digestive processing and once again puts it back in the cell. It may do this as much as a hundred times concentrating the nectar a little more each time until its sugar content is as high as 80%. And the result, of course, is honey.

Lots of animals love this incomparably sweet substance and are prepared to steal it from the manufacturers if they can. The bees' stings are powerful deterrents but even so, bears and chimpanzees, and even mice will risk getting badly stung in order to get a few mouthfuls of such delectable sweetness.

One African mammal has a particular addiction to it – the honey badger or ratel. And it has an ally in helping it find it – a bird, the honeyguide. It's about the size of a woodpecker, to which in fact it is related, and its taste is not, in fact, for honey but for the wax with which the bees construct their combs. But it has a problem. African bees often make their nests in crevices in rocks or holes in trees. Getting at it for a bird would either be very laborious or impossible. So the honey-guide seeks help from the honey badger, or indeed a human being. If you or a honey badger walks into its territory, the honey-guide soon appears and sits in a prominent position in a tree making a special rattling call. You walk towards it. It flies off and calls again. So you follow. Before long, almost certainly, you will see a hole in a rock or a tree with bees flying in and out. I've followed a honey-guide in this way and indeed shared the honey with my African companion after he had stupefied the bees with smoke from a burning log.

But what I haven't seen is the way a ratel does it. According to local stories it too has to calm the bees before taking their combs.

And it has a quite extraordinary way of doing so. It backs up against the bees' nest and rubs its backside against the entrance. Sometimes it will even do a handstand in order to do so. At the same time it twirls its tail and releases a strong smelling liquid from its anal glands which, it is said, knocks out the bees. However that may be, the ratel then pokes its head into the bees' nest and rakes out the combs with its claws. And having feasted on the honey, it leaves the empty combs around for the honeyguide to eat the wax. And it is traditional in Africa for a human honey hunter to also leave an empty comb in a prominent place to thank the bird for its guidance.

Human beings of course have raided bees' nests since antiquity. There are drawings on cliffs in southern Europe showing people doing so that are thought to be around ten thousand years old. And people today will take extraordinary risks in order to collect it. In southeast Asia the biggest of all bees, the giant bee, builds combs like immense dewlaps six feet across and suspends them from overhanging cliffs. In Nepal men climb up rickety ladders or descend on long ropes to get at these honey-loaded nests, taking appalling risks. In Borneo, this species of bee hangs its combs from the branches of giant trees. I was once hauled up alongside one of these in order to film it. It's quite an alarming experience because these bees, nearly an inch long, have a hugely powerful and painful sting. They cling to the surface of their comb covering it almost completely. If you get too near it, they lift their bodies in a coordinated way producing a kind of Mexican wave that moves across the face of the comb in rippling bands. I was wearing a bee-suit, of course, but I was told that giant bees had such huge stings they could sometimes pierce it. So I had no intention of really rousing them by taking any of their honey.

But I did get some in the local market. It's very expensive. People there gladly pay almost twice as much for it as they would for a jar of imported honey. And I can understand why. It is indescribably delicious.

So honey, or the nectar from which it comes, is relished by animals of all kinds. Mammals like me and the honey badger love it. In Australia, a little marsupial, the honey possum, has a special brush-tipped tongue to collect it from Banksia flowers. Sunbirds and hummingbirds live on virtually nothing else. Lizards lick it from blossoms. But, as I said earlier, honey appeared a hundred million years ago. So perhaps there were small dinosaurs that, like honey badgers, also relished it. It's quite possible. They were certainly around when it was first on offer. And I rather like that thought, as I spread honey on my breakfast toast.

*Pieter Bruegel the Elder drew this picture of beekeepers collecting honey in 1568 but the technique had been used for many centuries before that. The hives, known as skeps, were made of woven straw and the spectre-like men wear wicker masks, to protect themselves from angry bees.*

*The hawkmoth* Xanthopan morgani praedicta *is so-called because Darwin, having seen this Madagascan orchid* Angraecum sesquipedale, *with foot-long nectar-tubes dangling from its flowers, predicted that there must be an insect with a drinking-tube to match. It was many years before the hawkmoth that is so equipped was actually discovered.*

*The African ratel, a relative of the badger, eats a wide range of food – fruit and carrion, scorpions and nestling birds. But its favoured food is honey and it will use its formidable digging powers to rip open a tree in order to reach it.*

Above. *The honeyguide does not eat honey. Its taste is for beeswax but it can seldom if ever open the nest of wild bees unaided. It has therefore developed a partnership with ratels and will guide one to a bees' nest that it knows. The ratel may then open the nest in order to get at the honey and so expose the waxy combs for the bird.*

Below. *Several kinds of ants in Australia, Mexico and the deserts of the western United States, store honey by feeding it to a special caste of workers who take in so much that their bodies swell to the size of peas. They hang from galleries in underground nests and give up their honey if a worker solicits it during the season when food is scarce. These belong to an Australian species.*

*In Nepal men climb or lower themselves down cliffs in order to collect honeycombs from the nests of the giant Asian bee. Danger comes not only from the risk of falling but of being attacked by clouds of angry bees, which have a particularly powerful sting.*

Above. *The giant Asian bee hangs its huge exposed combs beneath the branches of the tallest forest trees as well as from cliffs. Nonetheless, thieves manage to reach them - geckos and moths which sip the honey and the honey buzzard which takes, not the honey, but the bees' larvae.*

Right. *Bees collect nectar in their crops, but they also have baskets on their back legs which they fill with protein-rich pollen and take back to the nest and feed to their growing young and their*

The
Nondescript
See
The Wanderings
W539

# 18

# Waterton

Squire Waterton of Walton Hall, at the age of 82, regularly climbed the oak trees on his estate, at great speed and barefoot. He was double-jointed so that he could put his right foot on the top of his head and pick up things with his toes.

◁ *The enigmatic stuffed specimen that, as an engraving, served as the frontispiece of Waterton's account of his South American travels.*

Squire Waterton of Walton Hall, at the age of 82, regularly climbed the oak trees on his estate, at great speed and barefoot. He was double-jointed so that he could put his right foot on the top of his head and pick up things with his toes. When talking to visitors, he had a habit of standing on his head or swinging from a door frame. He dressed like a farm-labourer except on formal occasions when he wore a blue swallow-tailed coat with brass buttons, which was rather too small and at least a century out of fashion. His house, Walton Hall in the West Riding of Yorkshire, was large and spacious but downstairs its rooms were almost fully occupied by stuffed animals. – though I should not really have said 'stuffed'. That would have infuriated the Squire, for these specimens, all of which he had prepared himself using his own special method, were in fact hollow. He himself slept elsewhere – in the attic, on bare boards covered with an old threadbare cloak and with a block of mahogany as a pillow. He was – you might say – eccentric.

He was born in 1782 and he owed his fame, initially, to an extraordinary book that he wrote about his travels in what was then British Guiana and is now the independent country of Guyana. His family owned sugar plantations there on the coast, and in 1804, at the age of twenty-two, he went out to manage them. Eight years later, he was able to leave them and venture into the great forests. One of his objectives was to find the source of the lethal poison that the forest-living Amerindians used to tip their blow-pipe darts. They called it 'wourali', Europeans 'curare' and the active element in it is known today as strychnine. But Waterton also collected specimens of the animals he encountered, turning their skins on the spot into the hollow versions that were his unique speciality.

He made four of these journeys to South America over the next twelve years and the book he wrote about them is one of the oddest travel books I know. My first trip to the South American rain-forest was to Guyana. I spent three months there and when I returned I sought out Waterton's book. It is called *Wanderings in South America* and in its time, it was a best seller. I found half a dozen copies of it sitting on the shelf of a second-hand bookseller. One, however was much fatter than the rest. That, I discovered, was because it had bound into it a dozen or so handwritten letters from Waterton. It was the same price as all the rest so naturally that was the one I bought.

The letters were amazing. They attacked other naturalists with unrestrained vigour. Audubon, for example, the North American ornithologist who was visiting Britain at the time, exhibiting his magnificent engravings of American birds, should be horsewhipped, wrote Waterton, for the errors of natural history they contained.

The book itself is very odd reading. It is written in almost Biblical style with lots of 'thee's and 'thou's. Its pages are strewn with quotations from the Latin classics. He uses local Guyanese names for the animals he encountered, instead of English or scientific Latin ones – although he knows both of those perfectly well. There is little continuity in his narrative – it is more a series of disconnected snapshots. And although most travel books of that time were liberally illustrated with engravings, in this one there is only a single illustration – the frontispiece. It shows the head and shoulders of a heavily bearded creature, apparently but not certainly human, with hair running continuously from the top of its head, over the neck and the chest. It has curling lips, a prominent nose and a rather baffled expression. What it is, is not clear. The text only refers to it in the vaguest terms and it is captioned simply *A Nondescript.*

But the book is full of his adventures investigating the wild life of Guiana. He does that in the most alarmingly practical ways. He catches a huge boa constrictor by first punching it on the jaw and then tying its mouth shut with his braces. He is so anxious to discover the truth about vampire bats that he spends night after night with his bare leg hanging out of his hammock trying to tempt them to suck his blood – but with no success. At one point he decides to catch a caiman, the South American crocodile. But he won't allow his men to shoot or spear one. That would spoil it as a specimen. So he ties a hook to the end of a rope and baits it. A ten-foot caiman takes it and Waterton in his excitement rushes into the river, jumps on the animal's back as it thrashes about in the water and grabs its front legs, holding them as if they were a horse's reins, while his men haul him and the animal ashore. That story caused such a sensation that George Cruikshank, one of the most famous of the Victorian cartoonists, published an engraving of the event which, like the book itself, became a best-seller. Charles Dickens said that he was 'very partial ' to it. Even Charles Darwin consulted the author.

But Waterton deserves to be remembered not only as a writer and an eccentric, an adventurer and a naturalist. He was much more important than all those put together. He lived at a time – the middle of the 19th century – when the Industrial Revolution in this country, where after all it had started, was gathering momentum and driving all before it. Coal mines were opening up all over Yorkshire. Near Walton Hall factories of all kinds were being built. Some were boiling animal fat to make soap. Others were manufacturing sulphuric acid. Noxious smoke was belching from chimneys two hundred feet high. Leaves shrivelled on the trees and people choked from the fumes. Slag heaps and slums were everywhere.

Appalled, the eccentric Squire Waterton, when he came back from his last journey in 1821, started to build a wall around his estate to hold back the tide of pollution as far as he could. Constructing it took him five years. In parts it was 18 feet high. Within it, all animals were safe – except the brown rat which the Squire detested and which he regarded as a hateful alien. He had built this country's very first nature reserve, maybe even the first in the world in this modern sense.

He caught what foxes and badgers there were within and released them in the surrounding countryside though later he wondered whether he should have done. He gave sixpence to anyone who brought him a hedgehog so that it could be released in his park. Instead of felling ancient hollow oaks, he propped them up since they made such good places for birds to build their nests. He created a special sandbank for sand martins. He constructed a tower with sixty special nesting holes, each with a loose stone at the back so that he could clamber up inside the tower and look inside the nests to see how the occupants were getting on – a technique that wildlife film-makers today will recognise. The estate had a lake. On it sailed flocks of waterfowl and herons nested in the nearby trees. And the Squire spent his days, in his workman's clothes and often barefoot in all weathers, revelling in his estate.

But he was far from a hermit. He welcomed anyone from the surrounding towns who wanted to share these delights with him. They could come in for nothing provided that they brought neither a gun nor a dog. In one year, 18,000 visitors came. Perhaps most touching and perceptive of all, he regularly welcomed mentally disturbed people from what were then called 'asylums' to come and sit in his front room and watch waterfowl on the lake through his telescope. He believed that it calmed them.

There was no doubt that Charles Waterton had a volcanic temper, as I had discovered from his letters. But my heart warmed to him when I discovered the row that lay behind that enigmatic frontispiece to his great book, the one that he captioned *A Nondescript.* It represents one of his mounted specimens that he prepared with special care. The fur comes from a howler monkey's bottom, but Waterton with his unique taxidermic skills, has moulded it to resemble a man's face. Whose? Well, it seems that when he arrived in Liverpool docks at the end of his last expedition, a customs officer unexpectedly turned up and demanded to inspect all Waterton's specimens. The duty that had been imposed on them, he said, was far too low. So they were all impounded and only released after a hugely increased duty had been paid. So, according to one biographer, the Squire of Walton Hall moulded a howler monkey's bottom into the custom officer's likeness and used it as the frontispiece of his best-seller. What more satisfying insult could a knowledgeable naturalist devise.

*The Nondescript engraved.*

Above. *Waterton's account in his book of how he rode a caiman caused such a sensation that George Cruikshank made it the subject of one of his cartoons.*

Below. *In 1851, an American showman arrived in Leeds with 34 live rattlesnakes. He tried to demonstrate to an audience of scientific gentlemen the lethal powers of his reptiles by providing them with a live rabbit. He was, however, terrified of his charges and when one escaped it was Squire Waterton who stepped forward and picked it up by grasping it around the neck with his bare hand.*

EXPERIMENTS WITH RATTLE-SNAKES, AT LEEDS.

Above. *Waterton used his taxidermic skills to produce three dimensional allegories. This one, entitled 'Ole Mr Bull in Trouble', is made from the shell of an extraordinarily hairy tortoise which carries on its back a heavily spiked dragon-like creature with two bulky purses, one labelled National Debt and the other Eight Million Pounds. It is escorted by a gigantic millipede, two bloated toad-like creatures, one of which has wings, and a lizard holding its tail over its back like a lemur.*

Right. *Waterton was a devout Catholic and resented a Protestant from Hanover becoming King of England. He expressed his views in this taxidermic allegory that he labelled Martin Luther after his Fall.*

Below. *A hitherto unpublished pair of Waterton's taxidermic works. One, on the left, is an accurately prepared head of a saki monkey. The other, also from a saki, he has transformed into a human-like face by narrowing its mouth, sharpening its nose and giving it a chin.*

# 19

# Fireflies

Every Christmas, the Royal Institution in London mounts a short series of lectures especially for children. It is obligatory to include experiments and demonstrations and the more spectacular they are the better.

◁ *The British glow-worm is not a worm. It is a beetle. Only the female lights up as she sits in a hedgerow or a grassy verge. She does so in order to attract the winged males as they fly by. She produces her light by a chemical reaction in the last three segments of her abdomen. If she is picked up gently, she will continue to do so while she sits on your hand.*

Every Christmas, the Royal Institution in London mounts a short series of lectures especially for children. It is obligatory to include experiments and demonstrations and the more spectacular they are the better. One of the regular ones, often dragged in on the feeblest of excuses, involves two large flasks, each containing a colourless liquid. The lights are lowered. The lecturer takes one flask and pours its contents into the other – and suddenly the darkened theatre is filled with an eerie greenish glow so bright that you can see the smiles of astonishment and delight on the faces of the children sitting in the front rows.

And it *is* a stunning demonstration. The chemicals concerned are called luciferin and luciferaze and they were first identified in insects – glow-worms and fireflies. Until not so long ago there were arguments about how such insects created their light. Some people maintained that it could only be produced with the aid of phosphorescent bacteria. But ultimately, biochemists proved that the reaction was a purely chemical one by synthesising the two compounds concerned. Both are produced in the body of an individual insect. Luciferaze is a kind of enzyme. When it comes into contact with luciferin it causes it to oxidise and nearly all the energy produced by the reaction is then released as light.

It is a very efficient process. Compare that with the sort of household bulb we used before eco-friendly ones were introduced. Only about ten per cent of the energy they required produced light. The rest was lost as heat. Glow-worms, since they illuminate themselves with luciferin, don't warm up at all when they turn on their little lights in an English hedgerow during the summer.

It is only the female that produces the light – from the back end of her abdomen. Glow-worms are not worms but beetles. She is wingless and during the day hides in a little sanctuary in the ground. There she preys on slugs and snails, injecting them with poison that also liquefies their flesh. It may take her several days to consume an entire snail. She uses her light to attract a male, producing it in protracted pulses. The tip of her body becomes increasingly bright over a period of several seconds and then slowly fades to blackness before starting again. The males are somewhat smaller. They have wings and very large eyes. An hour or so after dark, they take to the air and start to cruise around, looking for the light of a female. When a male spots one, he lands beside her, clambers on to her – and she takes him pick-a-back to her little home.

There are two species of glow-worm in Britain, closely related but one much rarer than the other. In the world, however, there are over a hundred. One in the West Indies produces a light so bright that people are said to put them on their heads in order to read – and on their feet to find their way around at night. A South American species is called locally the 'railroad worm' because the male has one spot on his head that produces a brilliant red light and a line of smaller ones down each flank that produce a pale green.

Yet another insect, a fungus gnat, produces a light that is a brilliant electric blue. They live in one or two New Zealand caves, each lying in a little tube of transparent mucus slung horizontally from the cave roof, like a hammock. And each is surrounded by dangling threads of silk, loaded with beads of glue. Midges hatch in the stream that flows below along the rocky floor of the cave. Attracted by the lights, they fly up to the ceiling and inevitably collide with a glue-laden thread. The fungus gnat immediately detects the vibrations, turns off its light and wriggles half out of its tube. It grabs the quivering thread, hauls it up and eats the entangled midge. Filming this was not easy, because if we added the amount of light we would normally use, it would drown out the light produced by the gnat. But it proved to be possible – just.

Another spectacular light-producing insect is found in the Far East. This is another beetle. It is not related to the fungus gnat and only distantly to the British glow-worm though like them it also uses luciferin. It is remarkable because tens of thousands of them gather together and flash simultaneously.

We found a village on the coast of Malaysia where the people told us of a tree on the outer edge of a nearby mangrove swamp where the males congregate in such numbers that the flashing tree can be seen from a distance of about a quarter of a mile. It is so bright that fishermen from the village used it for navigation at night in much the same way as we would use a lighthouse.

Late that afternoon, before it got dark, we set off in a canoe through the swamp. As darkness fell we could detect scattered flashes among the mangroves through which we threaded our way. Every now and then, an individual male would fly through the air ahead of us, tracing his path with a long curving arc of flashes. By the time we reached the tree, it was fully dark. All its branches were outlined by lines of little lights like those trees in towns that are illuminated at Christmas – except that these lights were flickering. We reached one particularly large tree and tied ourselves to an overhanging branch. Slowly the confusion of flickering in the branches began to resolve itself into a more coordinated fashion until eventually they were all synchronised and the entire tree pulsed with light, its image reflected in the black waters beneath and backed by the glittering stars of the tropical night.

Lighthouses, of course, flash their light with a distinctive and identifiable frequency – two quick flashes and then a long pause, for example. That enables seamen to identify a particular light so that they know exactly where they are. And fireflies do the same thing – to show what species they are.

There are half a dozen species of this group of fire-flies in North America. We went to a meadow in New England where there were five of them, each with its own characteristic call-sign. Here both sexes flash. The males do so while they are in the air, cruising around in search of females. One species for example, flashes every half second in long bursts. Another produces two flashes, a second apart. A third flashes only once and sporadically.

The flightless females sit on the top of grass stems. The feminine responses given by all the different species are much the same – a single flash – though the timing varies. The female of one species, for example, will reply exactly two seconds after the enquiring flash of the male. When he sees that, he changes his flight path and comes down to land beside her. Such a single flash is not too difficult to imitate. I tried it with a small pen-light torch, flashing it on to my outstretched finger. And it worked. A male, flying by, flashed. I waited the statutory two seconds and gave my answering flash. And down he came to land on my finger tip.

But alas! – the simplicity of the required responses has led to trouble among fireflies. The males are not only smaller than the females but outnumber them by as much as fifty to one. So males are only too eager to respond to a receptive female if they find one. And that can be a big mistake.

The males of one kind, called *Photuris,* signal by a characteristic sequence of staccato flashes. The female's response is, as usual, a carefully timed single flash. But this female will also send an acceptance signal in reply to the flashes of males of other species as they cruise by. If a male of one of these glimpses her reply and is deluded enough to come down and land beside her, she grabs him with her jaws and eats him.

This is not an accident. The female *Photuris* can identify the call sign of the males of all five species and she modifies her response to match the code of whichever male happens to be passing at the time. She is so skilled at doing this – and does it so frequently – that males of other species form a significant part of her diet.

Some people maintain that there are no moral codes in the animal kingdom. Only human beings have such refined feelings. Well, whether that is true or not, it is certainly the case that some animals – female *Photuris* fireflies for one – tell lies.

*Dr Muffet, whose daughter sat on a tuffet but famously didn't like spiders, included this woodcut of a glow-worm in his* Theater of Insects *in 1658.*

Above. *The larvae of a kind of fungus gnat hang from the roof of a New Zealand cave in hammocks of slime. Their light attracts insects hatching from the stream below which fly up and are entrapped in dangling threads loaded with beads of mucus.*

Below. *A female glow-worm, having attracted the much smaller male with her light, now carries him back to her nest where they will copulate.*

*A time exposure reveals the flight-paths of male fireflies as they signal to females congregating in a nut tree in the Philippines.*

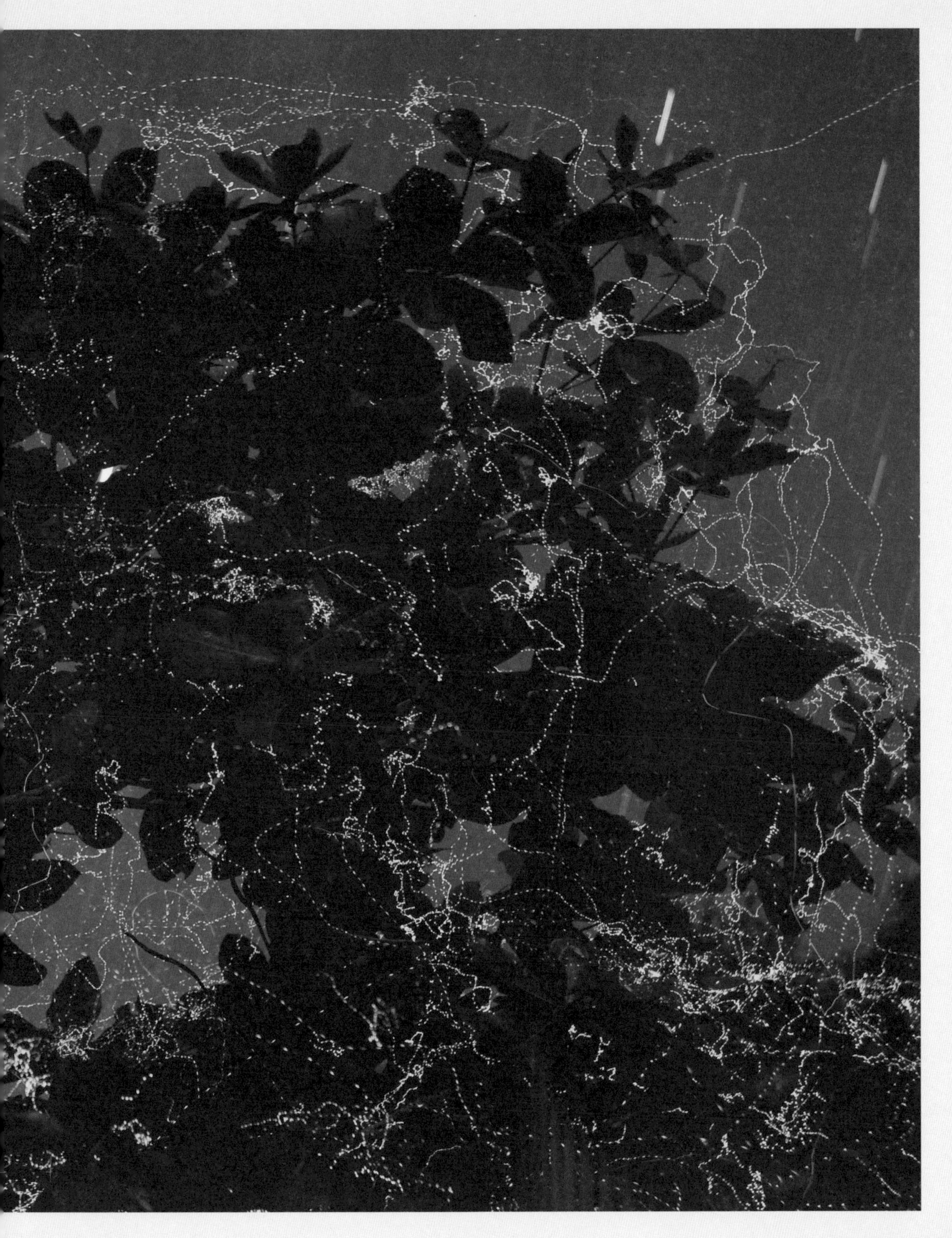

# 20

# Elsa

In the summer of 1960, the bookselling world was agog with excitement. A book was about to be published that was going to be a sure-fire best-seller. A sensation.

◁ *George Adamson greets Elsa and her cubs on the rocky summit of a kopje near the Adamsons' camp where she frequently basked during the day.*

In the summer of 1960, the bookselling world was agog with excitement. A book was about to be published that was going to be a sure-fire best-seller. A sensation. It was written by a woman who had hand-reared a young female lion cub. Nothing particularly sensational about that, of course, but this woman and her husband had also successfully returned the young lioness to the wild where she had mated. And then – and this was the exciting part – she had brought back her cubs to show them off to her human foster parents. The lioness's name was Elsa.

Cameraman Geoff Mulligan and I were about to leave for Madagascar when I was summoned by my boss, the Head of BBC Television's Talks Department. The Corporation had got exclusive permission to film Elsa – and Joy and George Adamson, the couple who had raised her. It was a great scoop. Would I therefore break my journey to Madagascar and go up to Northern Kenya and film what was happening.

When we landed at Nairobi I chartered a small plane and within a few hours, off we went. We landed on an improvised air-strip cleared in the bush close the Adamsons' camp. Joy was there to meet us – and she had bad news. Elsa the lioness had disappeared. For the past few days there had been signs of a strange lion around the camp and the previous night they had heard sounds of a ferocious fight in the bush nearby. Joy was certain that Elsa had been involved. Maybe she had been badly wounded – even killed. At any rate she had now vanished and we had no lion to film.

There was nothing to be done. The charter plane had already left and I was exhausted. So after a quick meal, I set up a camp bed in the shade of a huge, wide-spreading fig tree beside a lovely river – a paradisaical spot if ever I saw one – and went to sleep.

I was woken by a truly appalling stench of halitosis. I opened my eyes and saw – and the vision is still very clear in my memory – the underside of a huge jaw, its fur snaggled with saliva, just a few inches above my face. A lioness was leaning over me. Had I sat up with a jerk, I would have bumped into her chin.

While I was wondering what to do about this, I heard Joy's voice. 'Elsa, Elsa, mein Liebchen' she called. "Komm, komm.' – Joy, I should have mentioned, was an Austrian – and to my enormous relief the lioness heaved herself up and strolled off to the camp. Joy greeted her ecstatically, embracing and patting her as though she were an Alsatian dog. '*Jinja mbusi*' she called. For a moment I thought she was calling her lioness a ginger pussy. But no, she was speaking Swahili, telling her servants to kill a goat. So while lioness and lady continued their embraces in the front of the tent, the dreadful sound of a goat having its throat cut came from the back.

That night, Elsa's cubs came into camp. There were three – two females and a slightly larger more dominant male whom Joy called Jespah. They took a bit of getting used to. They were certainly playful, but equally, they didn't seem to know their own strength. Jespah in particular enjoyed playing games. His favourite trick was to hide behind a bush and then charge out as you were passing and take a swipe at your legs. Joy warned us that occasionally he neglected to withdraw his claws. That was why she always wore stout canvas leggings. The Africans working in the camp were very cautious about approaching him. So was I.

Joy told us how Elsa had come to them. George, who at that time was the senior game warden in this part of the country, had been called out to deal with a lioness that was raiding the cattle belonging to local people and which had attacked one of the herders.

After he had shot her, he heard a mewing noise coming from a hollow among some rocks and discovered three young cubs which he brought back to camp. Two he sent to European zoos. The third, Elsa, they kept.

Right at the beginning, however, he and Joy decided that they would return her to the wild just as soon as she was able to survive there. But that meant that they would have to teach her how to hunt – as her mother would have done. George did this by shooting an impala or some other animal, tying the body to a rope and then pulling it behind a car so that Elsa could practice pouncing on a moving target. In the wild, a lioness will often deliberately refrain from killing prey outright in order to enable her cubs to practice finishing it off. Whether George ever did that or not, Joy didn't say.

The film we made at the camp was not very good. Fifty years ago, we had no way of recording sound and filming pictures simultaneously, so we couldn't film an interview with either George or Joy. But it does contain shots of the real Elsa and they were the last to be taken, for a few months later, she died from tick fever.

Joy's book, entitled 'Born Free,' was a huge success, as everyone had predicted it would be and a feature film was produced that was based on it. To make it, George had to get more cubs and tame them so that it was safe for actors to appear with them. And that he did. And the film – like the book – was a huge success, with its theme tune 'Born Free' on everyone's lips.

But underlying it, as underlying the life of any lion, was violence. It had begun after all, when George had shot Elsa's mother. He then had had to kill innumerable antelopes to feed Elsa and to educate her as a hunter – let alone innumerable goats that had had their throats cut. But that was only part of it. After Elsa's death Jespah and his two sisters had wandered away from the Adamson's camp in search of territory and prey for themselves. But having been familiar with human beings from a very early age, they had no fear of them and the local game warden decided that they were dangerous to human beings and should be shot.

Before that could happen, George herded them into the back of his Land Rover and took them down to the Serengeti where he released them. Soon afterwards, a European on safari there and asleep in his tent was dragged out and killed by a lion. The game warden then followed its tracks and shot it. It was a young male. Whether it was Jespah or not, no one could be sure, but Jespah was never heard of again.

When the feature film of Born Free was made, George took on the job of providing several tame lions to play the parts of Elsa and her cubs. In the process however, one of them turned on George's assistant and killed him. So George had to shoot that lion too. Finally, Joy who had antagonised local people with her imperious ways, was murdered. George stayed on in his camp until he too was murdered by local bandits. Elsa the lioness may have been born free but her life was shot through with violence from beginning to end.

Natural history film-makers are sometimes accused of including far too many violent scenes. Deciding how much to include is often a difficult judgment. Too little and you risk turning what you hoped would be a serious and responsible film into a fairy story in which there is no pain or suffering and people and animals don't die but go gently to sleep. Too much and you yourself, let alone viewers, will find the sequences unwatchable. Sometimes when I have looked at what we have left on the cutting room floor, I think I have gone too far towards the first.

After one such film was televised, I received a letter from a lady who was appalled by it. 'It would be better,' she wrote, ' if you took all the money spent on making that film and used it to teach lions to eat grass.' Absurd? Of course. But on the other hand it represents one side of a paradox that has to be faced and accepted by anyone who makes natural history films – or, come to that, who tries to deal responsibly with the natural world – of which we, after all, are a part.

*There are a number of legends and fables about friendships between human beings and lions. This illustration comes from an edition of Aesop's fables that was printed in 1476.*

Above. *Elsa returned to camp after a fight with a rival lioness. She had a small wound in her side and allowed Joy Adamson to dress it with antiseptic ointment.*

Below *Joy greets Elsa as she was taking a siesta alongside George.*

Above. *Jespah, Elsa's male cub, carries away his meal, watched approvingly by his mother.*

Below. *The Adamsons' lions, which wandered freely into all of the tents, even chose, on occasion, to rest on the camp beds.*

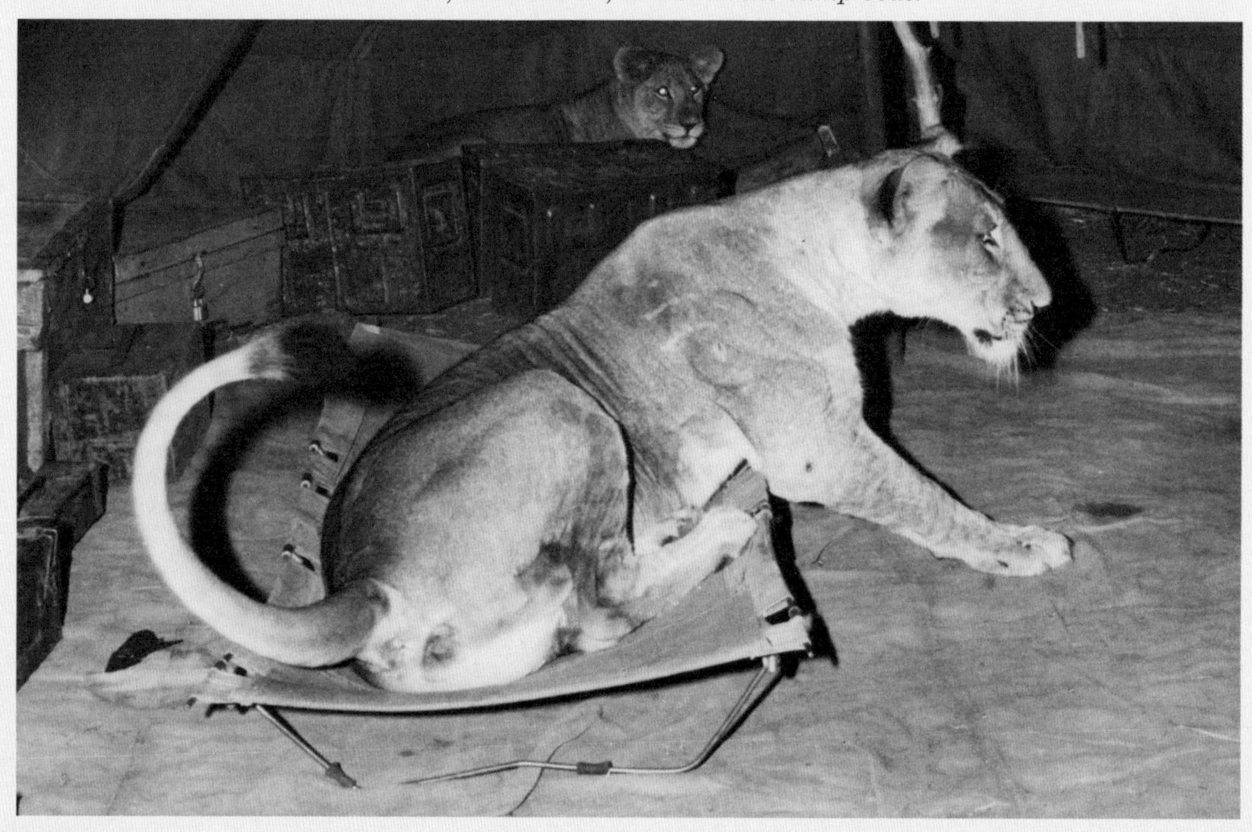

*Elsa occasionally decided to spend the night on the truck in which I was supposed to sleep.*

# Illustrations

Except for those listed below, all illustrations are from the author's personal collection or that of his friend, Mr Errol Fuller.

6 Mike Salisbury / naturepl.com
13 b Cyril Ruoso / Minden Pictures / flpa.co.uk
14-15 M. Watson / ardea.com
16 Thomas Marent / ardea.com
17 a Nick Gordon / naturepl.com
17 b Hans & Judy Beste / ardea.com
18 Pete & Judy Morrin / ardea.com
25 a Mark Boulton / ardea.com
25 b Rod Morris / rodmorris.co.nz
26 a Peter Bassett / naturepl.com
26 b Mark Cawardine / biosphoto.com
27 a Don Hadden / ardea.com
27 b Geoff Moon / flpa.co.uk
35 a Leicester Ciy Council
35 b Professor Guy Narbonne
37 a Professor Jim Gehling, South Australian Museum
37 b Professor Jim Gehling, South Australian Museum
38 b Professor Guy Narbonne
39 b Professor Guy Narbonne
47 b Nature Production / naturepl.com
47 c Inga Spence / flpa.co.uk
48 Natural History Museum, London
49 a François Savigny / naturepl.com
49 b M. Watson / ardea.com
50 Professor Gene Kritsky
56 The Royal Society
57 a Professor Gene Kritsky
57 b Professor Gene Kritsky
57 c Professor Gene Kritsky
58 a Professor Gene Kritsky
58 b Professor Gene Kritsky
59 a Professor Gene Kritsky
59 b Professor Gene Kritsky
60 Ralph & Daphne Keller / photoshot.com
67 a Kathie Atkinson / Auscape /ardea.com
67 b John Mason / ardea.com
76 b Cambridge University Library
78-9 Tim Laman / naturepl.com
88 a Bill Coster / photoshot.com
88 b Michael Fogden / OSF / photolibrary.com
89 James H. Hecht
90-1 Konrad Wothe / OSF / photolibrary.com
92 Frank & Joyce Burek / animalsanimals.com
98 corbis.com
99 François Gohier / ardea.com
100 Estate of Sir Peter Scott
101 a Franco Banfi / photoshot.com
101 b Patti Murray / animalsanimals.com
102-3 D.Parer & E. Parer-Cook / Auscape / ardea.com
104 Stephen Dalton / photoshot.com
110 Dennis Cameron / photolibrary.com
111 a Stephen Dalton / photoshot.com
111 b Stephen Dalton / photoshot.com
112 Joe Blossom / photoshotcom
113 Yvan Travert / photolibrary.com
114 flpa.co.uk
122 b Estate of Sir Peter Scott

| | |
|---|---|
| 124 | Ingo Arndt / Minden Pictures / flpa.co.uk |
| 131 b | Frans Lanting / flpa.co.uk |
| 132-3 | Ingo Arndt / Minden Pictures / flpa.co.uk |
| 134 | Kim Taylor / naturepl.com |
| 135 a | Ake Lindau / ardea.com |
| 135 b | John Mason / ardea.com |
| 136 | Stéphanie Meng / biosphoto.com |
| 143 a | Michel Gunther / biosphoto.com |
| 143 b | M. Watson / ardea.com |
| 144 a | Clive Bromhall / OSF / photolibrary.com |
| 144 b | Anup Shah / naturepl.com |
| 145 a | Ferrero Labat / ardea.com |
| 145 b | Anup Shah / naturepl.com |
| 146 | Frank Deschandol & Philippe Sabine / biosphoto.com |
| 153 a | Neil Bowman / flpa.co.uk |
| 153 b | Estate of Oliver Pike |
| 154 a | Yuri Shibnev / naturepl.com |
| 154 b | John Mason / ardea.com |
| 155 a | Paul Van Gaalen / ardea.com |
| 155 b | Paul Van Gaalen / ardea.com |
| 155 c | Paul Van Gaalen / ardea.com |
| 156 | Andrea Ferrari / photoshot.com |
| 164 a | Martin Dohrn |
| 164 b | Martin Dohrn |
| 165 a | Atlantic Productions Ltd / ZOO Ltd: created for the film *Flying Monsters 3D* |
| 165 b | Atlantic Productions Ltd |
| 166 | Joe McDonald / animalsanimals.com |
| 173 | Karl Terblanche / ardea.com |
| 174-5 | Thomas Marent / ardea.com |
| 176 a | Zig Leszczynski / animalsanimals.com |
| 176 b | Daniel Heuclin / biosphoto.com |
| 177 a | Mark Carwardine / ardea.com |
| 177 b | Zig Leszczynski / animalsanimals.com |
| 178 | John Mason / ardea.com |
| 184 | Bildarchiv Preussischer Kulturbesitz |
| 185 | Marcel Lecoufle / orchidée-lecoufle.com |
| 186 | Anthony Bannister / photoshot.com |
| 187 a | Nigel Dennis / photoshot.com |
| 187 b | Nature Production / naturepl.com |
| 188 | David Shale / naturepl.com |
| 189 a | Tim Laman / naturepl.com |
| 189 b | Kim Taylor / naturepl.com |
| 190 | Wakefield City Council |
| 196 | Wakefield City Council |
| 197 b | Wakefield City Council |
| 198 a | Wakefield City Council |
| 198 b | Wakefield City Council |
| 199 a | Wakefield City Council |
| 200 | Brian Bevan / ardea.com |
| 207 a | ANT / photoshot.com |
| 207 b | Robert Canis / flpa.co.uk |
| 208-9 | Jurgen Freund / naturepl.com |

# Index

Abominable Snowman, 123
Adamson, George, 210-219
  Joy, 210-219
Adelaide, 32
Aesop, 216
Africa, 21
Alaska, 85, 88
Aldrovandus, 172
Algeria, 168
Amazon, 71
Amerindians, 192
Andes, 47, 88
Angola, 115
*Angraecum sesquipedale,* 185
Anning, Mary, 158
Antarctica, 63
ants, 181, 187
Arctic, 96
artichoke, globe, 41
  Jerusalem, 41
Aru Islands, 71, 75
Audubon, John James, 94, 98, 193
Australia, 21, 32, 62, 181
Aztecs, 160, 162

Bahamas, 116
Baltimore, 51, 55
Banksia, 184
bat, short-tailed, 23, 27
  vampire, 194
batatas, 43
Bates, Walter, 70
bears, 182
bees, 178, 182
  giant Asian, 183-8
beetles, 71, 202
Belem, 71
Besler, Basilius, 41
Bewick, Thomas, 152
Bible, 52
Bikaner, 110
bindweed, 43, 47
birds of paradise, 6, 13, 69-73, 75, 77
  greater, 11, 75
  standardwing, 78
Bird, Dickie, 11
blackbird, 92, 93
boa constrictor, 126, 193
Boesch, Christophe, 139
*Born Free*, 214
Borneo, 71, 183
Brazil, 12, 160
Bruegel, Peter the Elder, 184
butterflies, 71, 125-35
  cabbage white, 128-9, 134-5
  heliconid, 131
  monarch, 124, 127, 131-2
  painted ladies, 128
  peacock, 128
  red admiral, 128
  swallowtail, 126
  tortoiseshell, 128

caiman, 194, 197
Cambrian, 30, 32
chameleons, 166-77, 173-7
Chance, Edgar, 149
Charnias, 29-39
*Charnia masoni*, 28, 31
Charnwood Forest, 29, 31-2, 34-5
chimpanzees, 94-5, 136-45, 182
cicadas, 50-9
Collinson, Peter, 56
Columbus, 42
Congo, 139
convolvulus, 43
coot, 26
Corner, Professor E. J., 8
Costa Rica, 90
Cousteau, Captain, 116
Crippen, Dr., 44
Cruikshank, George, 194
Cuba, 82
cuckoo, 146-55
Culpepper, Dr., 44
curare, 192
Cuvier, Georges, 158, 163

Darwin, Charles, 29, 63, 66, 72, 194
Darwinism, 74
Dawson, J.W., 34
de Montfort, Denys, 115
Dickens, Charles, 194
*Dickinsonia*, 37
dinosaurs, 21, 159
Dioscorides, 48
Dobo, 77
Down House, 76
Dragons, 116
dunnock, 150, 154

earthworms, 60-7
Ediacara, 36
Ediacara hills, 32
Egypt, 52
Eichstatt, 41, 163
elephants, 95
Elsa, 210-19

emus, 2
Everest, 119

fireflies, 200-9
Flay, Kevin, 97
fossils, 28-9, 30-36, 38-9, 119, 121, 158-9, 163, 180
*Fractofusus*, 38
fungus gnat, 203, 207
Gehling, Jim, 36
*Gigantopithecus*, 120
*Gigantosaurus*, 161
Gippsland, 62, 66
glow-worms, 200, 202, 207
Goodall, Jane, 94, 139, 143
Gosse, Philip, 121
Gould, John, 81-2
grasshoppers, 27
Great Exhibition of 1851, 82
Great Highland Fault, 116
Great Lakes, 127
Guyana, 192

Halmahera, 73
hawkmoth, 185
heliconia, 84-5
honey, 184
honey badger, 182
honey possum, 184
honeyguide, 182, 187
Hong Kong, 119
Hooker, Joseph, 72
hummingbirds, 80-87, 126, 184
  bee, 82
  copper-belted, 82
  puffleg, 82
  fiery-tailed awlbill, 82
  fiery-throated, 90
  green violetear, 88
  hook-billed hermit, 82
  marvellous spatuletail, 80, 83, 89
  rufous, 85, 88
  sapphire-spangled emerald, 82
  sparkling violetear, 82
  white-whiskered hermit, 82

Ice Age, 63
ichthyosaurs, 116, 158
India, 108, 110
Indonesia, 69
Industrial Revolution, 194

Java, 71
Jehovah, 52
jumars, 10
Jurassic, 31

kakapo, 22, 26
kangaroo, tree, 17
Kew, 72
kiwi, 18-27
Kritsky, Gene, 50, 57-9

Langston, Wann, 160
Lawson, Doug, 159, 165
Leicester, 31
Linnean Society, 72
lions, 96, 99, 210-19
lizards, 71, 184
Loch Ness, 116, 121
locusts, 52
London Zoo, 73, 171
luciferaze, 202
luciferin, 202
*Lumbricus terrestris*, 63
Lyell, Charles, 72
Lyme Regis, 158

macaws, 14-15
Madagascar, 170, 172
magnolias, 180
Magnus, Olaus, 120
malaria, 71
Malay Archipelago, 70
Malaysia, 204
mandrake, 44, 48
Mantel, Gideon, 158
Maori, 25
Mason, Roger, 31
meadow pipit, 149
mermaids, 116
Mexico, 85, 88, 127
mice, 182
Michigan, 127
milkweed, 127
Minnesota, 127
Mistaken Point, 32, 39
moa, 22-3, 25
Moluccas, 78
monkeys, 139
  colobus, 140, 145
  pig-tailed macaque, 13
  saki, 199
  spider, 16
Montfort, Denys de, 115
Moss, Cynthia, 95
Muffet, Dr, 206
Mulligan, Geoff, 212

narwhal, 116
Narbonne, Guy, 38
Natural History Museum, London, 117
nectar, 179-89
Nepal, 183, 188
*Nessiterras rhombopteryx*, 118, 122
New England, 52, 204
New Guinea, 11, 71, 74
New Zealand, 20, 203
Newfoundland, 33, 35

Ontario, 127

orchid, 181
Oregon, 85
Orinoco, 9
ostrich, 21, 24
Owen, Dr, 159
owls, 20

Packer, Craig, 96
Paraguay, 126, 131
parrots, 22, 26
penguins, 24, 97
 king, 102-3
Pennsylvania, 94
peppers, 45
 red, 49
 New World, 45
Peru, 49
Philippines, 230
*Photuris*, 205
Pike, Oiver, 149, 153
Pilgrim Fathers, 52
plesiosaurs, 116, 121, 158
Plymouth Rock, 52
Polynesians, 22
Portugal, 172
potato, 41
 sweet, 41, 43, 47
 Virginian, 46
Precambrian, 29-31, 32, 38, 39
*Pteranodon*, 164
pterodactyl, 158
pterosaurs, 159

Quetzalcoatl, 160, 162
*Quetzalcoatlus*, 156-65

ratel, 182, 186
rattlesnake, 197
rats, 105, 107, 109, 111, 113
 brown, 107, 194
 black, 107, 112
rheas, 21
Rines, Dr Robert, 117, 122
Rio Negro, 71
Rommel, 168
Rothschild, Lord Walter, 130
Royal Institution, 202
Russia, 160

Scott, Dafila, 96
Scott, Sir Peter, 96, 117, 122
*Semioptera wallacii*, 78-9
Serengeti, 215
shearwaters, 20
Shipton, Eric, 119, 123
Sicily, 172
Slimbridge, 96
snakes, 71
*Solanum*, 44
Solomon Islands, 106
Somerset, 62
soup, Palestine, 43
South America, 21, 43
Spain, 172
Spaniards Bay, 35
Spice Islands, 71
spiders, 97
 bolas, 97, 101
*Spriggina*, 37
squid, giant, 115-16
Stonehenge, 66
strychnine, 192
Stuart, Lachlan, 121
sunbirds, 184
swallow, 148
swan, mute, 96
 Bewick's, 96, 100
takahe, 22, 26
Ternate, 71
Texas, 159
tomato, 46, 49
toucan, 17
Topsell, Edward, 142
*Tyrannosaurus*, 161

unicorn, 116

Venezuela, 9
von Koenigswald, Ralph, 119

Waigeo, 76
Wallace, Alfred Russel, 69-79
Walton Hall, 192
*Wanderings in South America, 193*
warbler, reed, 146
Waterton, Squire, 191-9
weaver-birds, 150
West Indies, 43
Westminster Gazette, 86
weta, 23, 27
whale, grey, 95
 humpback, 95, 101
Wildfowl Trust, 96
Winkworth, Steven, 164
wourali, 192
wren, 148

*Xanthopan morgani praedicta*, 185

yeti, 119
Yorkshire, 192